GEOGRAPHY: THEORY IN PRACTICE
BOOK TWO

AGRICULTURE

RICHARD HUGGETT AND
IAIN MEYER

HARPER & ROW, PUBLISHERS
LONDON

Cambridge
Hagerstown
Philadelphia
New York

San Francisco
Mexico City
Sao Paulo
Sydney

1817

To Jane, Jamie, Sarah, and Edward
   Jacqueline and Lucinda

Copyright © 1980 Richard Huggett and Iain Meyer

Harper & Row Ltd
28 Tavistock Street
London WC2E 7PN

British Library Cataloguing in Publication Data

Huggett, Richard
   Agriculture. – (Geography, theory in practice;
   book 2).
   1. Agricultural geography
   I. Title   II. Meyer, Iain   III. Series
   338.1   S439

ISBN 0 06 318166 5

Typeset by Santype International Ltd, Salisbury, Wilts.
Printed and bound by Whistable Litho, Kent

GEOGRAPHY: THEORY IN PRACTICE
BOOK TWO
AGRICULTURE

Contents

Preface   **vii**

---

1   **The Economic Geography of Agriculture**   **1**

Introduction   2

   Net income and economic rent   5

   Economic rent and Von Thünen   6

   Von Thünen's theory in the modern world   8

Land-Use Intensity   12

   Economic rent and intensity of land use   13

   Down on the farm: local patterns   15

Land-Use Competition   19

   The formation of land-use rings   19

   Land-use rings in today's world   25

   Land-use rings distorted   39

Agricultural Regions   42

   Enterprise combinations in England and Wales   43

   Finding the right combination   48

Agricultural Change   52

   The spread of agricultural practices   52

   Supply, demand, and agricultural change   61

   Changes in farm structure   66

   Brittany: a case study   69

   Land reform in Italy   69

   Land reform in Bolivia   71

2   The Physical Geography of Agriculture   **72**

Climatic Limitations   73

   Soil moisture deficits in England and Wales   75

   Irrigation   80

   A case study: the Sudan   80

   The relationship between land use and altitude around Louth   81

Soil Limitations   87

   Land use and land capability near Eastnor, Herefordshire   89

   Soil type and agricultural yield in the Roe valley   96

The Pattern of Farming in Britain   98

   Tillage   99

   Grassland, cattle, and sheep   108

Biological Limitations   115

   Ecological efficiency   117

3   The Environmental Impact of Agriculture   120

The Impact on Soils   121

   Soil erosion by water   122

   Soil erosion in eastern North America   125

   Wind erosion   129

   The loss of soil structure   130

The Impact of Pests and Their Control   134

Preface

This book outlines and illustrates the chief theories and concepts of the geography of agriculture. These theories and concepts are explained concisely and, as far as possible, in jargon-free language. By means of carefully designed exercises, students should see how these theories work in practice. The exercises are, in the main, based on maps and tables adapted from a wide range of books and articles or on data from official sources. Worked examples of the techniques used are included, and carefully posed questions encourage students to interpret and to criticize their results. At the end of each chapter is a list of books which might be helpful in answering the questions or in essay work. Though written as a course-book, *Agriculture* has plenty of scope for teachers to bring in their own examples and sets of data.

Chapter One considers the economic geography of agriculture and in particular the intensity of land use, competition between land uses, and changes in land use. Chapter Two considers the physical geography of agriculture, in particular the limitations imposed by climate, soil, and biology. Chapter Three discusses the impact of agriculture on the environment, topics covered including soil erosion.

Richard Huggett   **Iain Meyer**
Macclesfield   Elstree

December 1979

CHAPTER ONE
THE ECONOMIC GEOGRAPHY OF AGRICULTURE

Introduction

Unlike their wild counterparts, farm crops and farm animals cannot be left untended. Cereals need sowing, weeding, and harvesting. Cows need feeding, milking, and veterinary care. A farmer cannot sit back and let nature take its course. He must acquire tools and machinery, prepare land, keep buildings, hedges, fences, and ditches in good repair, and look after his crops and stock. He must decide on what crops to grow, what animals to raise, and how much labour to take on. For his land to give a harvest which may be sold at a market, a farmer has to spend money on time, effort, and materials: to get something out of the land, something has to be put into it (Figure 1.1).

Farm inputs are land, labour, capital, and management. Land factors, which include size of farm, natural soil quality, slope, drainage, climate, and air, are natural attributes of the area in which a farm has been built. Many of these inputs can be manipulated by a farmer: soil quality may be changed by adding fertilizers, liming, and marling; a steep slope can be terraced; wet patches of land can be drained; to a limited extent, clouds can be seeded to induce rainfall or to avoid the harmful effects of hailstorms; the adverse effects of strong winds can be lessened by the prudent placing of rows of trees and hedges which act to break the force of the wind. Labour is the effort a farmer and hired hands put into the process of agricultural production – tilling, sowing, harvesting, milking, or whatever. Capital is all the farm inputs which have accrued over the years – tools, machinery, fences, land drainage systems, buildings, animal stock, and cash in the bank (or under the mattress). Management, or enterprise, is the process by which a farmer decides what to grow, how to grow it, and in which fields to grow it; in short, what types and what quantities of input to use.

3

Inputs

Farmer and his farm

Outputs

MANAGEMENT

CAPITAL
Farm buildings,
money in bank, animal
stock, etc.

LABOUR

LAND
Size of farm, natural
soil quality, slope,
drainage, climate.

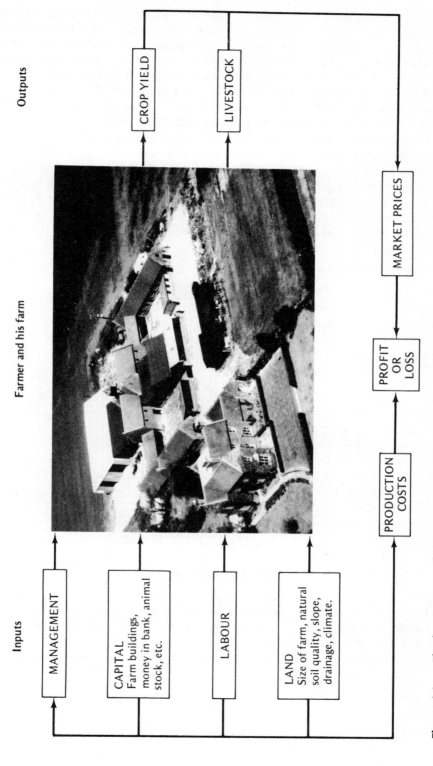

CROP YIELD

LIVESTOCK

MARKET PRICES

PROFIT
OR
LOSS

PRODUCTION
COSTS

Figure 1.1   The farm system. Photo reproduced with permission from Aerofilms.

4

Farm output is the quantity or tonnage of yield from all agricultural enterprises on a farm, both crops and livestock. It is not the same thing as the value of the output, otherwise called the income. The output of a particular farm may stay more or less the same year after year, but the income made by selling the output might vary from one year to the next because of the vagaries of the agricultural market, some products being in short supply one year but forming a glut in another.

Farm outputs vary with the kind and quantity of farm inputs. As the quantity of a given input, say fertilizer, is increased, so the output, say the yield of wheat, will increase, to start with anyway. Figure 1.2 shows that the addition of one bag of fertilizer to a hectare of land raises the wheat yield by half a tonne; the addition of a second bag of fertilizer raises the wheat yield by a further 0.4 tonnes. Each extra bag of fertilizer raises the wheat yield less and less until a peak output of wheat is reached at five bags of fertilizer. Putting down a sixth bag of fertilizer would actually reduce the output of wheat – too much can be as bad as too little. The overall relation between input and output shown in Figure 1.2 is an example of the law of diminishing returns.

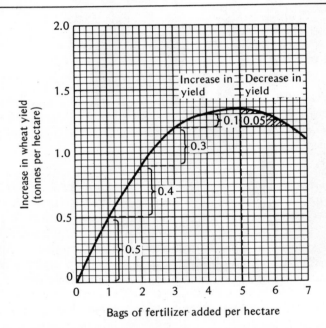

Figure 1.2   The law of diminishing returns applied to fertilizer inputs and wheat yields. The more fertilizer that is put on a field, the higher the yield of wheat. But the extra wheat yield from each additional bag of fertilizer gets smaller and smaller until, with the addition of a sixth bag of fertilizer, the yield actually decreases.

NET INCOME AND ECONOMIC RENT

The annual accounts for a 300-hectare farm are listed in Table 1.1.

Table 1.1 A yearly account for a farm

Expenditure		Income (from sale of products)		
		Enterprise	Area	Value of output
Production costs per hectare	£50	Winter wheat	40 ha	£4400
		Winter barley	30 ha	£2800
		Winter oats	30 ha	£2800
Total expenditure on production (50 × 300)	£15,000	Sheep	100 ha	£4000
		Beef cattle	100 ha	£7400
				£21,400

Net income = Gross income − Expenditure
= £21,400 − £15,000
= £6400

The production costs are £50 per hectare, giving a total of £15,000 for the 300 hectares of the farm; this would include charges for labour, land improvement and maintenance, the cost of seed, and so forth. The value of the farm output is £21,400. So the net income for the year is £21,400 less £15,000, or £6400.

An adjacent farm of the same size, with the same production costs, and the same proportion of land given over to the same five enterprises, could well earn a higher net income because the land on which it lies is naturally more fertile. Within any region, farms on the least fertile land are said to be at the economic margin of cultivation. Other things being equal, all farms on more fertile soil in the region will receive a higher net income than farms at the economic margin of cultivation. The difference between the net income of a farm on fertile land and the net income of a farm at the economic margin of cultivation is called economic rent. Imagine two farms growing wheat, one of which lies in a naturally more fertile area than the other. The farmer working the fertile land has the same production costs as the farmer working the less fertile land but, because his land yields more wheat per hectare, he earns economic rent; the farmer on the economic margin of cultivation earns no economic rent. So economic rent of a parcel of land is the additional income above the income that can be obtained from a parcel of land at the economic margin of cultivation. Now assume both farmers are tenants on an estate, that is, they rent their land from a landlord. The farmer working the fertile land can use his economic rent to offer more rent for his land than the other farmer – better land yields higher economic rent and therefore commands higher rents (Figure 1.3). This view of economic rent and its relation to soil fertility was first taken by the economist David Ricardo in the nineteenth century.

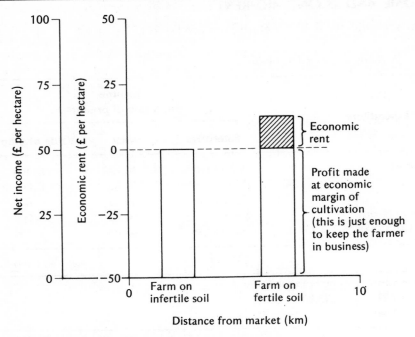

Figure 1.3   Economic rent as described by David Ricardo

ECONOMIC RENT AND VON THÜNEN

David Ricardo considered differences in economic rent resulting from variations in soil fertility. In 1826, his contemporary, Johann Heinrich Von Thünen, produced a far more comprehensive analysis in his book *The Isolated State*. He considered such factors as long-run crop rotations, changes in technology, and distance from market, as well as soil fertility. It is Von Thünen's in-depth study of the distance-to-market factor which has been popular with geographers. Von Thünen noted, from his many years' experience managing an estate near Rostow near the Baltic Sea, the tendency for economic rent for a crop to fall as the distance to market increases – the further from market, the greater the costs of transporting produce to sell on market day, and so the less the economic rent.

Imagine two farms, one rented by Mr. Hoe, the other by Mr. Rake. Both farms grow barley and get five tonnes on every hectare of land. Production costs, in which we shall include the sum needed just to stay in business, for every hectare of barley are £60 on both farms. Both farmers receive £20 for a tonne of barley so they have an annual gross income per hectare of £100 and an annual economic rent per hectare of £40. However, Mr. Hoe's farm is one kilometre from the local market and Mr. Rake's farm is forty kilometres from it. The going rate for transporting barley to market is 20 pence per tonne per kilometre; so the entire output from one hectare – five tonnes – costs £1.00 per kilometre to transport to market. Mr. Hoe's transport costs are £1.00; deducting the costs of transport from his economic rent leaves £39.00. On the other hand, poor Mr. Rake has to find £40.00 to pay for transport, which reduces his economic rent to zero – his production costs plus transport costs just equal the value of his produce at market, and he is operating at the economic margin of cultivation for barley (Figure 1.4). Farmer Rake gets no economic rent but Farmer Hoe earns £39's worth. Because in Von Thünen's analysis economic rent drops with distance from market, some workers have styled it locational rent; Von Thünen himself used the term *Bodenrente* – land rent.

1   Construct a graph (Figure 1.4) to show how economic rent declines with increasing distance away from the market in the region in which Hoe's and Rake's farms lie. Assume that there is a farm every two kilometres, and that, before deducting transport costs at a rate of £1.00 per hectare of production per kilometre, all farms receive an economic rent of £40.00 per hectare.

What happens to a farmer growing barley forty-two kilometres from market?

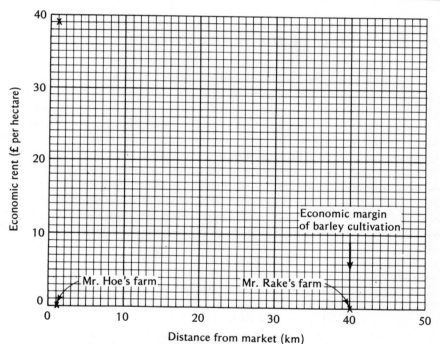

Figure 1.4   Economic rent as described by Von Thünen

In singling out the effect of distance-to-market on economic rent, Von Thünen made several assumptions about other factors which influence economic rent. Some of these assumptions were:
i   one mode of transport for shipping produce to market – in Von Thünen's days this was horse and cart;
ii   all farms could transport produce to market with equal ease and at the same cost

for a unit of production (tonne, cow, or whatever) per kilometre;

iii all farmers were so-called economic men who would try to obtain the largest possible returns from their land; and

iv production costs were everywhere the same.

More of Von Thünen's assumptions are contained in the following extract from the translation of his book:

Imagine a very large town, at the centre of a fertile plain which is crossed by no navigable river or canal. Throughout the plain the soil is capable of cultivation and of the same fertility. Far from the town, the plain turns into an uncultivated wilderness which cuts off all communication between this State and the outside world.

There are no other towns on the plain. The central town must therefore supply the rural areas with all manufactured products, and in return it will obtain all its provisions from the surrounding countryside.

2 Having read the extract, list the assumptions stated in it.

VON THÜNEN'S THEORY IN THE MODERN WORLD

A body of theory concerned with agricultural location has grown out of Von Thünen's ideas. It stresses the links between the farm, or centre of production, and the market, or centre of consumption, farming patterns being related to the cost of transport between the two. In essence, Von Thünen argued that the proportion of a farmer's total costs spent on transporting produce to a central urban market increased with distance from the market. Different enterprises had different transport costs and so different economic rents. The land was used to take advantage of the varying levels of economic rent which could be gained from different enterprises. From these premises, Von Thünen derived a land-use pattern in which bulky and perishable goods were located close to the central urban market, while other enterprises were placed in a series of concentric rings around the market. The inputs required by each of the enterprises declined from the market in each successive zone until uncultivated land was reached.

The socio-economic conditions to which Von Thünen applied his analysis – limited transport by horse and cart, self-contained production areas and market areas (isolated states), and a free market economy – are characteristic of a bygone age. Von Thünen's theories did account well for the land-use patterns of his day. The Reverend Henry Hunter, a contemporary of Von Thünen's, had observed during his travels in the London region four zones of land use around the city: on the edge of the built-up area were clay pits, the clay being used in the making of bricks; outside this was an extensive belt of pastureland supporting dairy cattle; along the banks of the River Thames, between Bow and Hampton, was a wedge of market gardening; this was followed by an outer zone in which hay was produced. Livestock, as Von Thünen was soon to show, could be reared far from the market; it was not unusual for Welsh cattle to be sold in London having been fattened on the pastures in the South East. A land-use survey of the London area made in 1800 by Thomas Milne also revealed a pattern of land use which closely followed Von Thünen's model.

But changing technology has dated the application of Von Thünian principles, to some enterprises at least. Costs of transport have declined in relative terms and today represent a mere 5 per cent or so of total costs. The diminishing importance of transport costs has meant that physical factors have tended to play a more important role in the fashioning of agricultural patterns. In Britain, horticulture used to be carried out on the outskirts of most towns to provide the population with fresh fruit and vegetables. The advent of railways in particular, and the improvement in transport in general, has led to the large-scale production of horticultural crops in areas where climate and soil favour high yields, at the expense of small-scale production by market gardeners in areas around towns which may not be physically very suitable for growing horticultural crops. Transport is still important in determining the locational pattern of some enterprises: this is the case for bulky goods like potatoes and sugar beet, and also for vegetables which are quick-frozen or canned. Pea- and sugar beet-processing factories are markets which create supply zones around them similar to the supply zone around the central city in Von Thünen's isolated state (Figure 1.5a and b). H. D. Watts (1974), in a study of the British sugar beet industry, has shown that the cost of taking beet to a beet processing factory increases away from the factory (Figure 1.5c). Since Britain's entry into the European Community and the placing of the sugar beet industry on a more rational footing, beet cultivation in Hampshire, Sussex, and Pembrokeshire, areas remote from the main processing plants in East Anglia, Lincolnshire, Yorkshire, and the West Midlands, has waned. The supply area of each processing plant has contracted to reduce transport costs.

R. T. Dalton (1971) has shown that, prior to 1965, the vining of peas more than 32 kilometres from the processing factory at Grimsby was unprofitable. Indeed, as transport costs had to be met by the farmer, peas tended to be vined within 16 kilometres of the factory. In 1965, the introduction of mobile viners reduced transport costs and this change in technology extended the economic margin of pea production.

Von Thünen's principles, then, can still be used in explaining the geography of some enterprises. But to what extent can they be used to explain present-day land-use patterns? To answer this question we shall look at two important aspects of agricultural geography – land-use intensity and land-use competition.

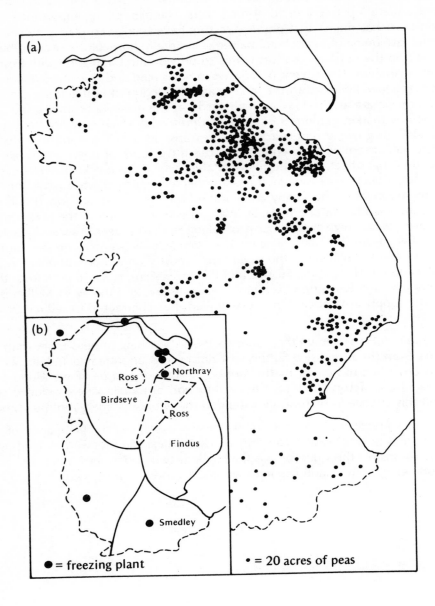

Figure 1.5 Pea cultivation in Lincolnshire, 1962
(a) Peas grown for freezing in 1962
(b) Areas covered by main companies and the location of freezing plants
Reprinted with permission from 'Peas for freezing – a recent development in Lincolnshire agriculture' by R. T. Dalton (1971). *East Midlands Geographer*, 5(3), 133–141, figure 1.

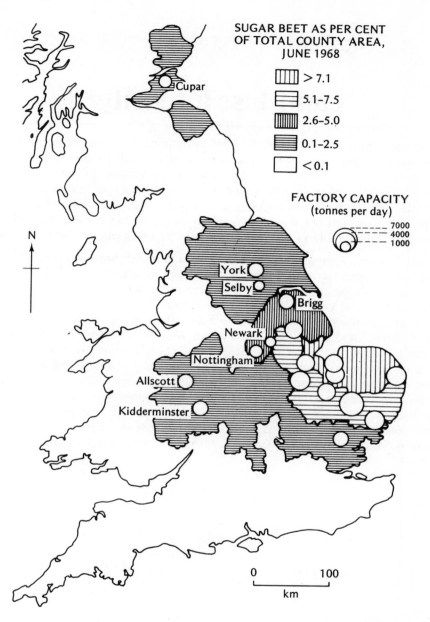

Figure 1.5c   Density of sugar beet cultivation by counties in 1968 and capacity of sugar beet factories in 1971. Circles refer to factory capacity. Names refer to towns with factories near spatial margins of profitability in 1971.

Reprinted with permission from 'Locational adjustment in the British sugar beet industry' by H. D. Watts (1974). *Geography*, 59.

Land-Use Intensity

Different farmers may decide to use different combinations of inputs in the production of the same crop. The inputs for a unit area of farmland, say for one hectare, define the intensity of land use. One farmer may buy bags of fertilizer in the hope of obtaining a bumper yield of wheat; a second farmer may simply rely on the natural fertility of the soil. The first farmer would be using the land more intensively than the second. So intensive land use is a large quantity of inputs per hectare of land; nonintensive or extensive land use is a small quantity of inputs per hectare of land.

One of Von Thünen's findings was that crops and livestock tend to be more intensively farmed (receive more inputs per hectare) the nearer the farm is to the market. On the large estate studied by Von Thünen, intensive agriculture – the production of milk and vegetables on heavily manured fields – formed the innermost ring of land use: in the early nineteenth century, when transport was slow and refrigerated lorries unheard of, perishable goods were best grown near the market. Forests occupied the next outer zone: large amounts of timber, a bulky crop and costly to transport, were used in the market as a building material and a fuel. The next three land-use zones displayed a successively more extensive form of agriculture, the dominant land use in each being rye and potatoes, rye, and rye with animal products: by lowering production costs through less intensive cultivation, the enterprises bore higher transport costs as the market became more and more distant. The outermost land-use zone was livestock farming or ranching; beyond this was wasteland.

ECONOMIC RENT AND INTENSITY OF LAND USE

The decrease in land-use intensity away from a market can be derived from a consideration of economic rent and distance. By way of example, take two potato farms run by Mr. Hay and Mr. Stack. Mr. Hay's farm is ten kilometres from market; Mr. Stack's farm is thirty-five kilometres from market (Figure 1.6). Both farmers have the same production costs per hectare of land so Mr. Hay's economic rent, because he has the lower transport costs, is greater than Mr. Stack's. Mr. Hay can therefore afford to spend more money on inputs, and so increase his yield of potatoes, than can Mr. Stack. Let us say Mr. Hay decides to spend money on spray irrigation equipment. Table 1.2 lists the costs of potato farming for intensive (with irrigation) and extensive (without irrigation) cases.

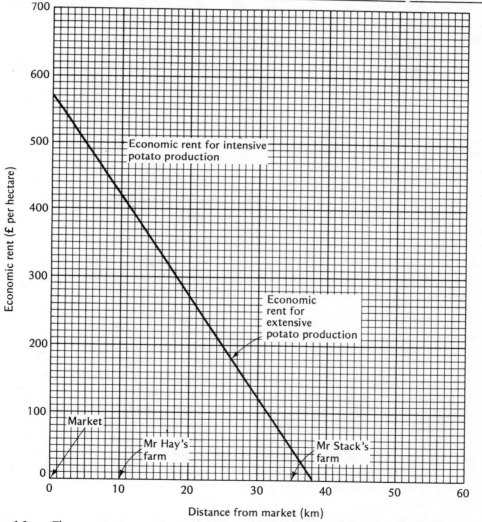

Figure 1.6   The economic rent from intensive and extensive potato farming around a market

Table 1.2 The cost of potato farming

Farm	Intensity of production	Distance from market (km)	Production costs (£/ha)	Yield (tonnes/ha)	Market price (£/tonne)	Transport rate (£/tonne/km)	Total transport costs (yield × distance × transport rate) (£/ha)	Total costs (production plus transport) (£/ha)	Gross income (yield × market price) (£/ha)	Economic rent (gross income less total costs) (£/ha)
Mr. Hay's	Intensive	10	100	40	20	0.50				
	Extensive		30	30	20	0.50	150	180	600	420
Mr. Stack's	Intensive	35	100	40	20	0.50	700	800	800	0
	Extensive		30	30	20	0.50	525	555	600	45

1a Complete Table 1.2 by calculating the economic rent for intensive potato production on Mr. Hay's farm. Mark the economic rent for both farms on Figure 1.6 and connect the points for intensive potato production.

b Assuming that both farmers will try to make as much money as possible, what would be the pattern of land-use intensity around the market?

c At what distance from the market is the economic margin of potato cultivation?

d How would the pattern of land-use intensity change if the production costs of intensive potato farming were to be cut by 25 per cent, assuming that no matter how many potatoes are produced they will always be sold at market?

Does the pattern of land-use intensity described and derived by Von Thünen apply today? August Lösch, writing in the 1940s, argued that because the cost of some inputs (wages and actual rent for instance) decreases with distance from a market, and because technological advances have led to a general, relative decline in transport costs, intensity of land use may actually increase away from a market. R. Sinclair, in a paper published in 1967, showed that a similar effect is produced by urban sprawl: as land near cities is likely to be built on, farmers are unwilling to invest much capital in it and put it under nonintensive cultivation. Because of this, economic rent increases away from the market, for a little distance at least. A study of land-use changes around Rockford, Illinois, has vindicated Sinclair's idea. But there is an important situation where the intensity of land use does still accord with Von Thünen's theory. This is the case where the people who work the land live in a central settlement, be it village or farm, and travel to the fields each day. Under these circumstances, the distance between fields and farmhouse or village influences the use, and intensity of use, of farmland, a fact noted by Von Thünen.

DOWN ON THE FARM: LOCAL PATTERNS

Several studies, well summarized by Michael Chisholm in his book *Rural Settlement and Land Use*, have revealed that of the many factors which are influenced by distance from a farmer's home, the key factor is the time taken to commute between farmhouse and field. The longer it takes to get to a field, the more the cost of labour — commuting costs have to be added to production costs. The tendency is for more distant fields to be put under enterprises with undemanding labour requirements, and for nearer fields to be given over to labour-demanding enterprises. High labour inputs give rise to intensive farming, low labour inputs to extensive farming: in general then, intensity of production declines away from a farmer's home.

Chisholm has shown how this process works around Canicatti, a Sicilian village. Using land-use maps, Chisholm found that, apart from a small amount of irrigated land and land growing citrus fruits, suitable areas for which are very limited, the pattern of land use is as follows: around the village is an inner zone of intensive tree and herbaceous crop cultivation — vegetables, olives, almonds, and vines; beyond this is a zone devoted to cereal production and pasture. The percentage of land devoted to a variety of crops in successive distance zones around Canicatti, as well as the annual labour requirements per hectare of crop, is shown in Figure 1.7. Notice how vines, a labour-demanding crop, are found closest to the village, whereas unirrigated arable land and pasture, both undemanding of labour, are found furthest from the village. Other examples which demonstrate the same principle abound. Birot and Dresch have described the zoning of land use around villages in Bulgaria in the days before the Communist government came to power and introduced the collective-farm system. The land-use zones were vegetables, fruits, some vines, and cotton on plots of land by farms; then, very close by the village, communal pastureland and woodlots, followed by labour-demanding crops — vines, roses, and tobacco. Outside this zone were arable crops such as wheat and maize grown in a two-year rotation system. On the fringe of the village land were a few pioneer rose gardens and vineyards making inroads on the forest. In northern Nigeria, the pattern of land use around a village was found by Mansell Prothero to be like this: within the village walls are well-manured 'gardens' growing vegetables and culinary herbs used on a day-to-day basis; next, up to a distance of one kilometre outside the walls, land fertilized with manure provides continuous crops — guinea corn, cotton, tobacco, and groundnuts; the next 0.8 to 1.6 kilometres are occupied by unmanured land which is cultivated for three to four years, then allowed to revert to bush for at least five years to regain its fertility; and fourthly is a zone of small clearings in heavy bush.

Piers Blaikie studied the pattern of agriculture around the Indian village of Daiikera, 30 kilometres from Jodphur, in western Rajasthan. Apart from a small area of land irrigated by wells, the land is unirrigated and relies for water on a meagre and unreliable monsoon. The major part of the agricultural work has to be carried out over a short period of time, a migration to the fields from the village taking place at the coming of the monsoon. The rates for the hire of labourers and bullocks (for draught purposes) are high. Access to fields is by a network of dirt tracks used by traditional bullock carts which radiates from the village. Fields lying between two tracks are often

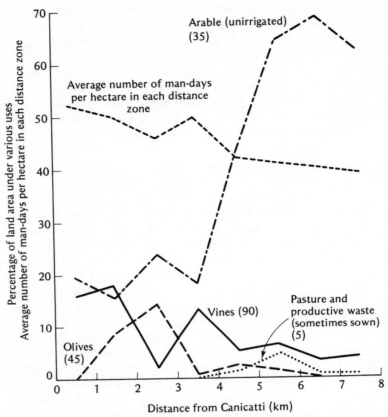

Figure 1.7 The percentage of land under selected uses and the average number of man-days per hectare in eight distance zones around the village of Canicatti, Sicily. The figures in brackets are the average number of man-days per hectare devoted to each enterprise. *Source of data: Rural Settlement and Land Use* (1st edition) by M. D. I. Chisholm (1962) published by Hutchinson, London, table 6.

difficult to get to. Cattle are reared in the village and manure taken to fields by cart. A farmer will have several fields scattered throughout the region; this compensates for the unpredictability of the monsoon rain which may well, in any one year, fall on some of a farmer's fields but not others. Blaikie costed the outlay on transport for each crop by timing porters, carts, and camels per thousand yards both loaded and unloaded, and multiplying the times thus obtained by the current seasonal rates for hire for human and draught animal labour. The cost, in man-days, of clearing, manuring, ploughing, levelling, sowing, weeding, bunching of the crop's leaves (in the case of sugar cane), harvesting, and threshing or crushing or podding for each acre of crop was worked out by timing how long the process took per acre and then multiplying this by the number of times each process was carried out for each crop.

1 Figure 1.8a shows land-use zones around Daiikera and the increase in labour requirements per thousand yards from the village per acre of land. Figure 1.8b shows the mean labour inputs per acre around Daiikera. Describe and explain the relation between crop zones and labour requirements.

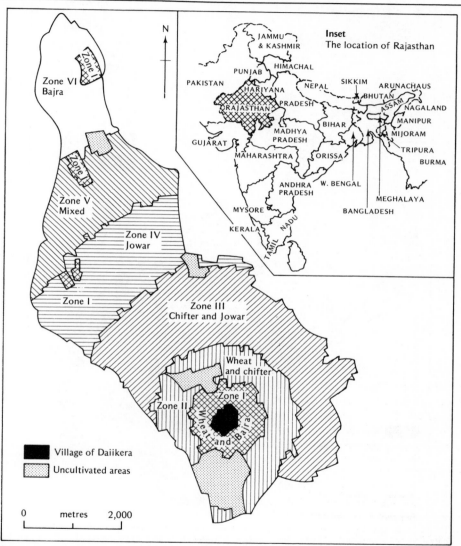

Figure 1.8a   Crop zones around the village of Daiikera, India. The increase in labour requirements (man-days per acre) for each crop per thousand yards from a farmer's home are: wheat, 28.3; chifter, a sort of millet cut fresh every day for fodder, 16.2; jowar and bajra, both types of millet, 12.6 and 10.9 respectively.
Reprinted with permission from 'Spatial organization of agriculture in some north Indian villages. Part I' by P. M. Blaikie (1971) *Transactions* of the Institute of British Geographers, 52, 1–40, figure 3.

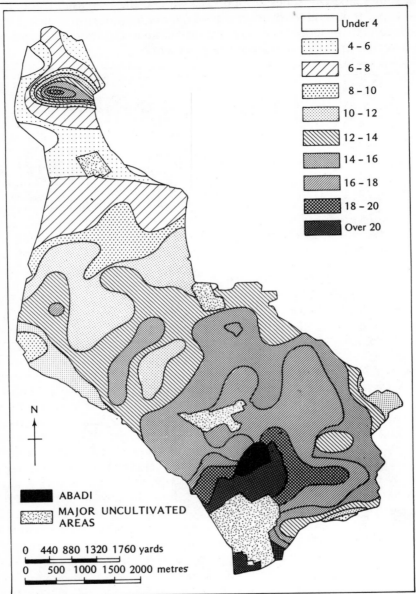

Legend:
- Under 4
- 4 – 6
- 6 – 8
- 8 – 10
- 10 – 12
- 12 – 14
- 14 – 16
- 16 – 18
- 18 – 20
- Over 20

N

ABADI

MAJOR UNCULTIVATED AREAS

0　440　880　1320　1760 yards

0　500　1000　1500　2000 metres

Figure 1.8b　Mean labour inputs (daily wage equivalents) per acre around the village of Daiikera.
Reprinted with permission from 'Spatial organization of agriculture in some north Indian villages. Part I' by P. M. Blaikie (1971). *Transactions* of the Institute of British Geographers, 52, 1–40, figure 8.

2　What other information would be helpful in giving a full explanation of the land-use pattern around Daiikera? For instance, what effect might field size, tax on the land, amount of fallow land, and the whims of the cultivators have?

Land-Use Competition

THE FORMATION OF LAND-USE RINGS

In the example on p. 13 potatoes were the only crop cultivated – the farmers practised monoculture. In some areas, monoculture is carried out, the vineyards in parts of France being an example, but, by and large, farmers opt to grow a variety of crops or raise many sorts of livestock or, perhaps, both. Von Thünen's most well-known finding is that, where farmers mix enterprises, the pattern of land use will, other factors being constant, be dictated by the distance from market; this arises because each enterprise has its own production costs, yield, market price, and transport costs. In consequence, the economic rent for each enterprise on farms located just by the market, where cost of transport is negligible, will be largest for the enterprise which has the biggest difference between gross income, calculated as value of output per hectare (market price × yield), and production costs. The economic rent for each enterprise will be less, the further a farm is from the market. Because the cost of transporting produce to market varies with the type of product – it costs more per kilometre to transport one tonne of milk than it does to transport a tonne of wheat – the fall in economic rent with distance from market (as depicted in Figure 1.4 for barley) will vary from enterprise to enterprise. The economic rent for enterprises where the products are relatively expensive to transport – milk and vegetables for instance – will fall sharply with distance, so giving a steeply sloping line on an economic-rent–distance graph. Enterprises whose products are relatively cheap to transport, such as cereal growing and livestock raising, will fall more gradually with distance, so giving more gently sloping lines on an economic-rent–distance graph (Figure 1.9a). The differences in transport rates mean that the enterprise which gains the highest economic rent at the market does not necessarily have the highest economic rent at some distance away from the market. Indeed, several different enterprises may yield the highest economic rent in different distance zones around the market. In the hypothetical example portrayed in Figure 1.9a, vegetables yield the highest economic rent at the market, but, being a costly commodity to ship, the economic rent from vegetable production plunges quickly to the economic margin of vegetable cultivation just five kilometres from market. Dairying yields the next highest economic rent at market. Dairy products are expensive to transport, though not as expensive as vegetables – the economic rent for dairying falls a little less steeply than that for vegetables and the economic margin for dairying lies ten kilometres from market. Rye and sheep farming both receive low economic rents, even at the market; but rye grain and sheep are comparatively inexpensive to transport, sheep more so than rye: both show a decline in economic rent with distance from market, the decline for sheep being less than the decline for rye. The economic margin of rye cultivation lies twenty kilometres from market and that for sheep farming lies sixty kilometres away. In this case then, the most profitable enterprise a farmer can choose to carry out depends how far his farm is from the market. As indicated on Figure 1.9a, for farms within a zone up to three kilometres from the market, vegetables give the highest economic rent; for farms in a zone between three and seven kilometres from market, dairying is the most profitable enterprise.

1a Assuming farmers are out to make as much profit as possible, what will be the land-use pattern beyond seven kilometres from the market?

 b What special problems would be faced by farmers whose farms lie three, seven, and fifteen kilometres from the market?

Notice that in this analysis a series of concentric land-use zones – Von Thünen's rings of land use – is produced (Figure 1.9b) by what in effect is different enterprises competing for land. This, then, is the process called land-use competition.

To find the economic rent per hectare of land for an enterprise at any distance from market, the following formula may be used:

Economic rent = yield (market price – production costs)
           – (yield × transport rate × distance from market)

where suitable units of measurement are yield – tonnes per hectare; market price – £ per tonne; production costs – £ per hectare; transport rate – £ per tonne per kilometre; distance from market – kilometres. Table 1.3 lists hypothetical yields, prices, and costs for sheep and wheat farming.

Table 1.3

Enterprise	Yield (per ha)	Market price	Production costs	Transport rate*
Sheep	1 sheep	£60 per sheep	£40 per sheep	£0.20 per sheep per km
Wheat	5 tonnes	£20 per tonne	£12 per tonne	£0.20 per tonne per km

* Though the transport rates appear to be the same for both enterprises, they differ if expressed as a rate per hectare of output – one hectare produces one sheep and so the output from a hectare costs 20 pence to take to market; on the other hand, one hectare of land yields 5 tonnes of wheat which would cost a total of £1.00 to take to market.

To calculate economic rent from sheep farming at one kilometre from a market using the formula, we have

Economic rent = 1.0(60 − 40) − (1.0 × 0.20 × 1.0)
         = 20 − 0.20
         = £19.80

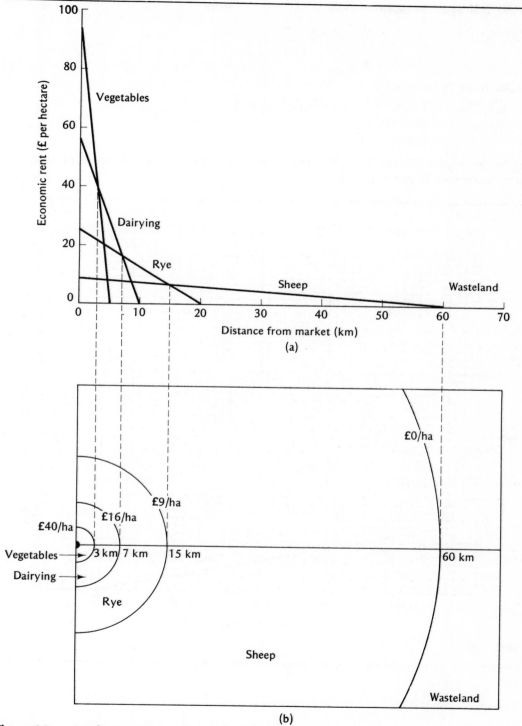

Figure 1.9   Land-use rings resulting from land-use competition

22

At ten kilometres from the market

Economic rent = 1.0(60 − 40) − (1.0 × 0.20 × 10.0)
  = 20 − 2
  = £18.00

And at ninety kilometres

Economic rent = 1.0(60 − 40) − (1.0 × 0.20 × 90.0)
  = 20 − 18
  = £2.00

Plotted on a graph (Figure 1.10), the results show the decline in economic rent from sheep farming with distance from market.

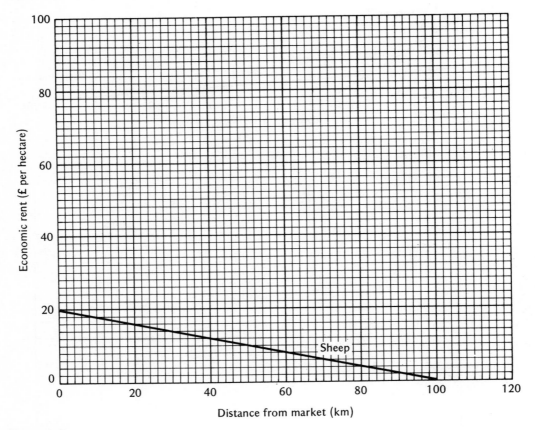

Figure 1.10

2a Using the economic rent formula and data in Table 1.3, calculate the economic rent from wheat growing at one, ten, and twenty kilometres from market.

 b Plot the results on Figure 1.10 and join the points by a straight line.

 c Describe and explain the likely land-use pattern in the region shown in Figure 1.10.

3 Figure 1.11a shows the market towns of Schnappsberg and Dachsundorf in a featureless plain of uniform soil fertility. Figure 1.11b shows the pattern of economic rent from three enterprises – dairying, potato farming, and sheep farming – in the region around the two markets.

 a Complete the outline map (Figure 1.11a) to show how land use around the fictitious German towns would likely be arranged assuming the farmers strive to earn the highest possible economic rents.

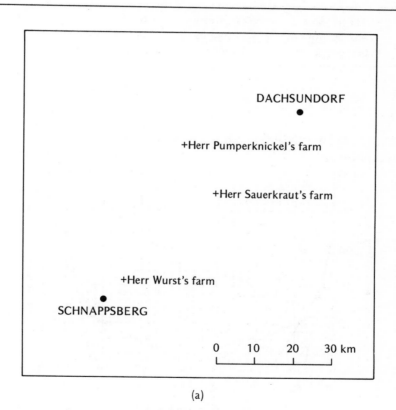

(a)

Figure 1.11a   A map of the Schnappsberg–Dachsundorf region

24

b   Using the data in Table 1.4, apply the economic rent formula to ascertain what would happen if the market price of all three types of produce were to double. Present your answer as a graph (Figure 1.11c).

c   On Figure 1.11c, indicate what would happen if the market price of potatoes were to double but dairy produce and sheep fetched the same market prices as in Table 1.4. Comment briefly on the situation which arises under these conditions.

Table 1.4   Annual costings for three enterprises in the Schnappsberg–Dachsundorf region

Enterprise	Production costs (DM)	Yield (per ha)	Market Price (DM)	Transport rate (per unit of production — litre, tonne, sheep per km)
Dairying	0.015 per litre	2000 litres	0.05 per litre	0.005 DM
Potatoes	0.462 per tonne	32.5 tonnes	2.00 per tonne	0.105 DM
Sheep	15.000 per sheep	1 sheep	45.00 per sheep	1.000 DM

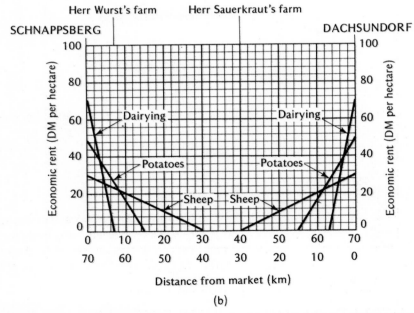

Figure 1.11b   A transect from Schnappsberg to Dachsundorf showing economic rent for three enterprises

SCHNAPPSBERG   DACHSUNDORF

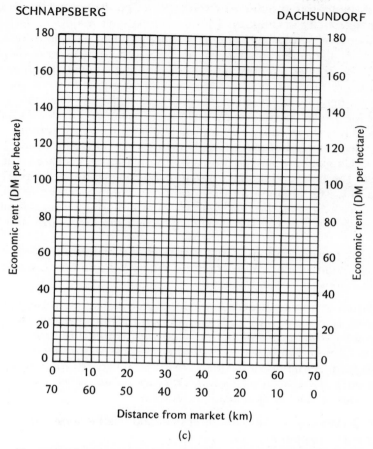

Distance from market (km)

(c)

Figure 1.11c   Transect of economic rent from Schnappsberg to Dachsundorf after market price for all enterprises has doubled

LAND-USE RINGS IN TODAY'S WORLD

The pattern of land use which in theory results from competition between enterprises – the concentric rings – did, as we have seen (p. 8), accord with land-use patterns in Von Thünen's time. But what of today's world? Are concentric zones of land use around markets still to be found in the agricultural landscape? In some areas, the answer to this question seems to be yes.

A case in point is the agricultural land-use pattern in Uruguay. A small nation with an area of 180,000 square kilometres, Uruguay more or less conforms to Von Thünen's notion of an isolated state: to the east is the Atlantic Ocean, to the south lies the Río de la Plata, to the west runs the Río Uruguay, the other side of which is Entre

Ríos – an undeveloped part of Argentina – and to the north lie lightly populated grazing lands of southern Brazil. Montevideo, with a population of 1.2 million, is far and away the biggest city in Uruguay. It is also the nation's major centre for imports and exports. Almost all surplus agricultural produce is taken to Montevideo for local consumption or for shipping abroad. Lying on the south coast, Montevideo is not, as it should be in Von Thünen's model agricultural landscape, located in the middle of the state. As a result, land-use rings in Uruguay can be expected to be truncated by the coast. The topography of Uruguay, consisting as it does in the main of undulating plains of slight local relief, is sufficiently uniform to meet Von Thünen's limitation of a flat area. Climate is uniform too, a subtropical régime with warm summers, mild winters, and an even spread of rain throughout the year prevailing over the entire country. Soils in Uruguay are not, as they are in Von Thünen's isolated state, uniformly fertile. The best land lies in a belt running roughly from Montevideo, along the southern and western margins of the country, to the border with Brazil. The poorest soils cover almost all the northwestern quarter and a big portion of the east-central part of the country. The remaining areas carry moderately fertile soils.

1 How do you think these variations in soil fertility might influence the intensity of agricultural production which, if Von Thünen's theory were to apply, would decrease evenly in all directions from Montevideo?

Von Thünen assumed primitive and evenly distributed transport. In Uruguay, the transport network is in some ways rudimentary but it is not uniformly dense or efficient. The road system in the south and southwest is well integrated and well maintained; the road system in the north and east is less well integrated and sparse. All major roads lead to Montevideo. There are no major routes running from east to west. Montevideo is also the focal point of a railway system which uses antiquated equipment and has inadequately maintained tracks.

2 In which parts of Uruguay would more intensive land use be favoured by a more efficient, and hence cheaper, transport network?

All in all, despite some variations in soil fertility and transport which, as we shall see, can be accommodated by relaxing some of Von Thünen's assumptions a little, Uruguay's physical and economic landscape is uniform enough to serve as a test site for Von Thünen's theory.

If the Uruguayan landscape complied with all the assumptions made by Von Thünen, we should expect to find Montevideo surrounded by four arcs or truncated rings of land use, something along the lines depicted in Figure 1.12a. But this pattern must be modified to allow for variations in soil fertility and transport facilities. So, modifying it, the land-use pattern can be expected to look like that drawn in Figure 1.12b, in which the rings of land use have been shifted a bit to the west.

3 Figure 1.12c is a map of agricultural regions in Uruguay. To what extent does the modified model landscape predicted by Von Thünen's theory (Figure 1.12b) match up to the actual land-use pattern (Figure 1.12c)? Make a note of both the similarities and the differences in your answer.

4a Why do you suppose that not all the land uses described by Von Thünen (see p. 12) are present in Uruguay?

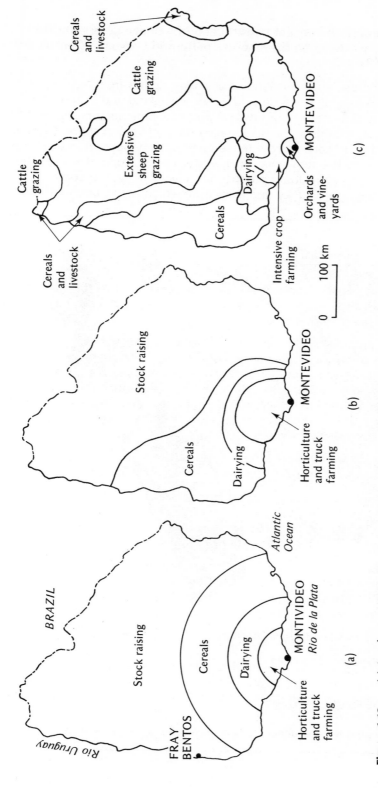

Figure 1.12 **(a)** Land-use rings in Uruguay which might be expected if all economic and physical factors in Uruguay complied with the assumptions of Von Thünen's ideal agricultural landscape

(b) Theoretical land-use rings in Uruguay modified to allow for areas of more fertile soils and cheaper, more efficient transport systems

(c) Actual agricultural land-use regions in Uruguay

(b) and **(c)** are adapted from 'Testing the Von Thünen theory in Uruguay' by Ernst Griffin (1973). *Geographical Review*, 53, 500–516.

5 What effect do you suppose the recent construction of a bridge across the Río Uruguay at Fray Bentos will have on the land-use pattern in Uruguay? (Consult an atlas.)

Another example of land-use rings surrounding a market comes from New South Wales, Australia, a state roughly three times the size of England and Wales. Here land use is fashioned by the interplay of physical and economic factors (Figure 1.13). The wet, rugged, infertile uplands are agriculturally unattractive and support forests and a few livestock farms; in contrast, the fertile coastal plains support arable farms and market gardening; and the somewhat drier western plains support sheep. The home market for the state's agricultural produce lies chiefly along the coast, especially around the towns of Sydney and Newcastle which house one-third of the

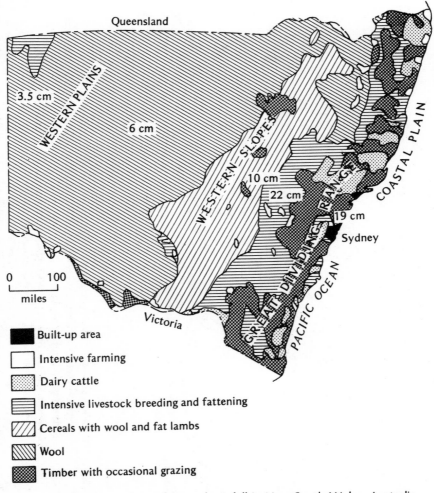

Figure 1.13   Land use, terrain, and annual rainfall in New South Wales, Australia

state's population. Sydney is also a major export terminal. Produce from much of the state has to be hauled vast distances to the home market and to be exported. We should thus expect Von Thünian principles to be in force; this does seem to be the case: Figure 1.14 shows how, by simplifying the land-use patterns in three easy stages, Von Thünen's rings, truncated by the sea, can be derived.

6a Examine Figure 1.14 and explain how the land use has been made more simple in each of the three steps.

b   Do you think the concentric-ring interpretation of land use in New South Wales is valid?

The general level of transport costs has fallen since the time Von Thünen was managing his estate. We should therefore expect land-use rings to expand and to

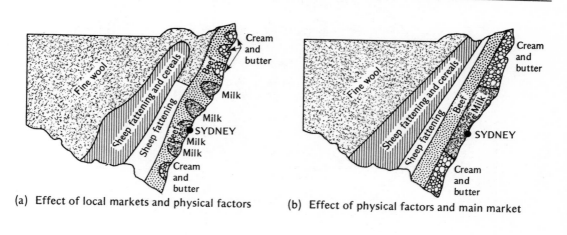

(a) Effect of local markets and physical factors

(b) Effect of physical factors and main market

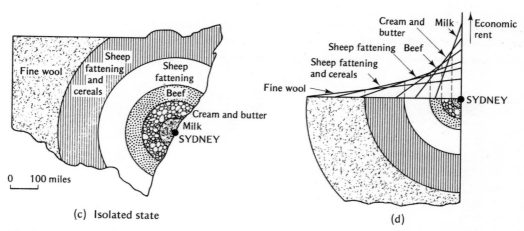

(c) Isolated state

(d)

Figure 1.14   Stages in a Von Thünian interpretation of land use around Sydney
Adapted from *New Viewpoints in Economic Geography* by J. Rutherford *et al.* (1966) published by Martindale Press, Sydney, figure 2.5.

move outwards. Where markets are closely packed the outer rings are pushed to the periphery of the state. The inner rings still serve individual markets but the outer rings now supply a collection of towns which form one large marketing centre (Figure 1.15). This seems to have happened in Europe where the countries in the northwest now form a vast central market: the annual application of fertilizer and yield of wheat decline from the northwestern core region to the periphery (Figures 1.16a and b).

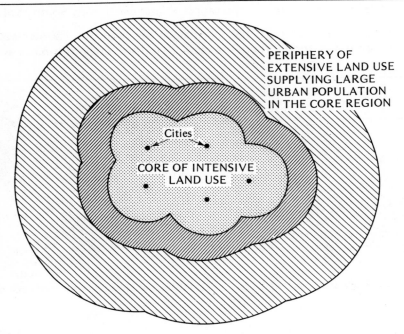

Figure 1.15 The fusing of land-use rings to form a core of intensive land uses and a periphery of extensive land uses

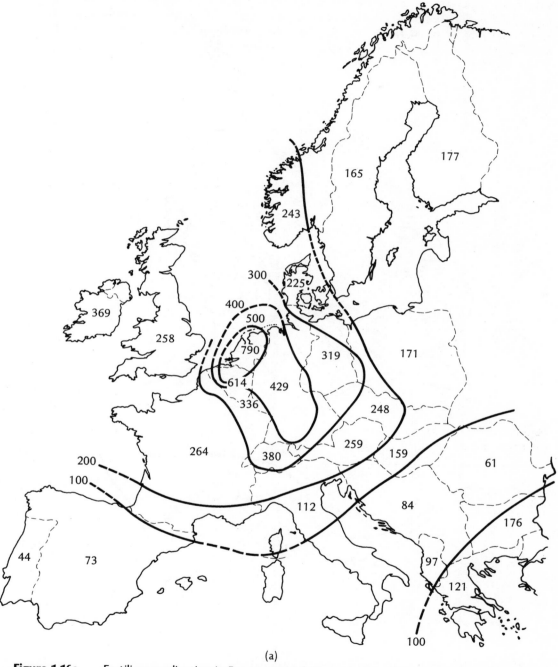

Figure 1.16a   Fertilizer application in Europe, 1971 (kilogrammes per hectare of cultivated land)

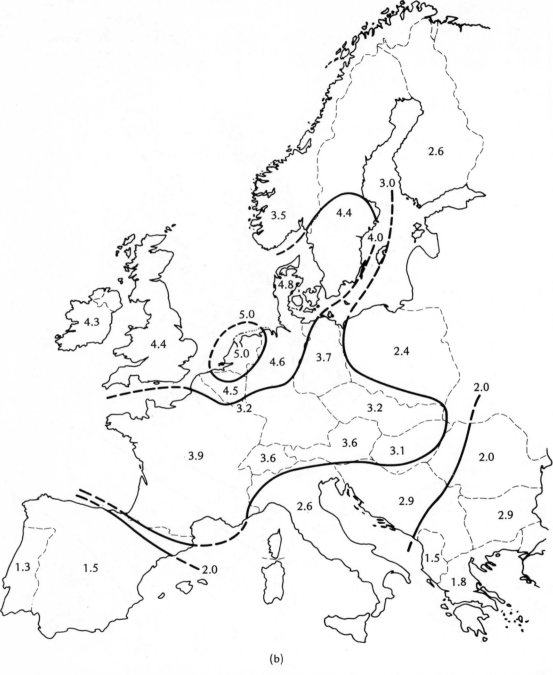

(b)

Figure 1.16b   Wheat yields in Europe, 1971 (tonnes per hectare)

7a   Using data in Table 1.5, construct an isopleth map for those states listed with a contour interval of $10 of average net income per acre from farmland in the United States (Figure 1.17c).

b   Using Figure 1.17b as a guide to potential agricultural markets, discuss how far the productivity of farmland in the United States (as reflected in net incomes per acre) and distance to markets are related.

Table 1.5

Selected States	Average net income per acre.* (US dollars)
Washington	30
Oregon	15
California	25
Nevada	8
Arizona	10
Montana	8
Wyoming	4
New Mexico	6
Texas	5
Oklahoma	10
Kansas	12
Nebraska	14
S. Dakota	15
N. Dakota	10
Minnesota	15
Maine	10
S. Carolina	20
Arkansas	15
Missouri	20
Iowa	22
Wisconsin	20
Michigan	25
New York	30
Indiana	30
N. Carolina	30
New Hampshire	30
Pennsylvania	42
Virginia	39
Connecticut	50
Delaware	50

* Net income per acre data are based on the results of Muller's trend surface analysis of county statistics.

The fusing of land-use rings as they grow outwards does not stop at the edge of states or even continents. The growth of London during the nineteenth century pushed the land-use zones outwards. Between 1850 and 1900, the rise in the standard

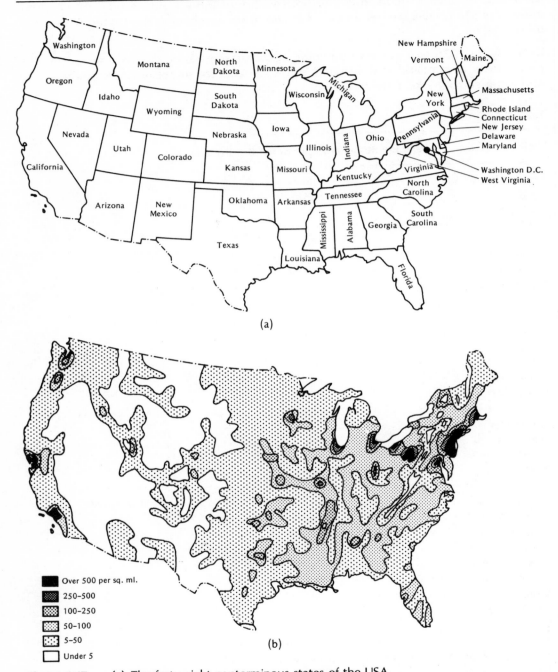

Figure 1.17   **(a)** The forty-eight conterminous states of the USA
                        (b) The pattern of population density in the USA
(b) is reprinted with permission from *A Social Geography of the United States* by J. Wreford Watson (1979) published by Longman, London and New York, figure 1.2.

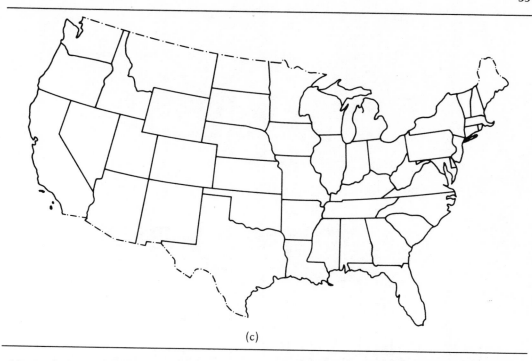

(c)

of living led to increased demand for meat, dairy products, and fruit. Farmers working land near the city found that cheese and butter gave a better profit than grain. And, as inland and ocean transport improved, so more semiperishable food was imported.

8 Using data from Table 1.6, construct graphs (Figure 1.18) to show the expansion of London's agricultural sphere of influence between 1833 and 1911. Label each curve and, with the aid of an economic atlas, attempt to explain the pattern of spread.

Table 1.6

Commodity imported	Average distance from London to point of supply (miles)				
	1833	1858	1873	1893	1911
Fruit and vegetables	0	324	535	1150	1880
Live animals	0	630	870	3530	4500
Butter, cheese, eggs	262	530	1340	1610	3120
Feed grains	860	2030	2430	3240	4830
Flax and seeds	1520	3250	2770	4080	3900
Meat and tallow	2000	2900	3740	5050	6250
Wheat and flour	2430	2170	4200	5150	5950
Wool and hides	2530	8830	10,000	11,070	10,900

Reprinted with permission, from 'The spatial expansion of commercial agriculture in the nineteenth century: a Von Thünen interpretation' by J. R. Peet (1969). *Economic Geography*, 45(4), 283–301, table 1.

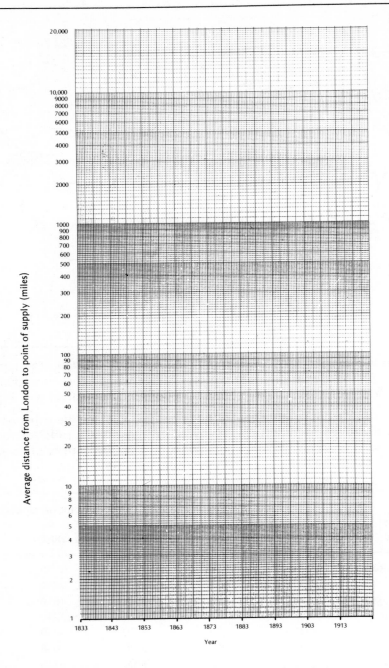

Figure 1.18 Average distance from London to point of supply (miles)

Note: This figure uses a logarithmic scale on the *y* axis. This differs from the graph paper with which you are most familiar since this scale increases at a geometric rate (whereas the *x* axis increases at an arithmetic rate), that is, it increases at a constant rate of ten and is divided into cycles, the end of each cycle being ten times greater than the end of the previous one.

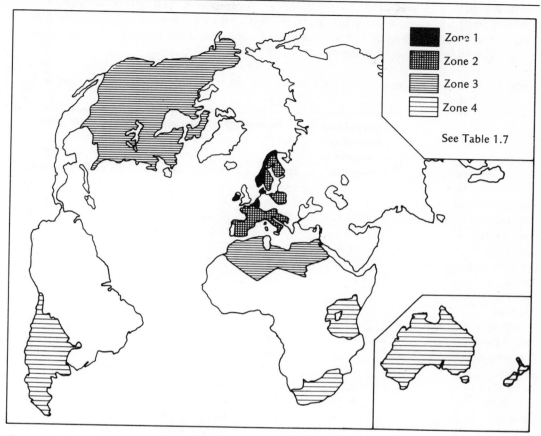

Figure 1.19 Imports of selected horticultural and dairy products in the United Kingdom, 1960 to 1962. The oblique azimuthal projection shows distances correct from London. Reprinted with permission from *Rural Settlement and Land Use* (1st edition) by M. D. I. Chisholm (1962) published by Hutchinson, London, figure 6.

At present, the eastern seaboard of the United States and the core region of northwestern Europe in effect form a gigantic agricultural market around which lies a series of international supply zones.

9 Examine Figure 1.19 and Table 1.7 which show the United Kingdom's imports of selected horticultural and dairy products divided into four zones of origin for 1960 to 1962. Using your newly acquired knowledge of agricultural economics, try to explain the geographical pattern of imports.

Table 1.7 United Kingdom, 1960–1962: imports of selected horticultural and dairy products (by weight)

Commodity	Imports as percentage of gross supplies	Percentage of imports originating from zone:				
		Zone 1	Zone 2	Zone 3	Zone 4	Other countries
Potatoes, not new	0.9	67.0	—	22.8	—	10.2
Cabbages	1.1	93.6	—	—	—	6.4
Strawberries	2.3	4.4	85.7	2.8	—	7.1
Carrots	6.8	25.6	15.0	50.4	—	9.0
Lettuces, endives	9.0	88.1	—	—	—	11.9
Plums	10.8	1.0	72.3	0.4	14.1	12.2
Asparagus	11.1	—	74.6	—	—	25.4
Cherries	11.3	—	94.7	2.8	—	2.5
Cauliflower, broccoli	11.3	2.2	95.9	—	—	1.9
Cucumbers	16.4	82.8	—	14.2	—	3.0
Potatoes, new	29.8	11.9	38.8	44.9	—	4.4
Apples	30.9	2.5	15.8	19.3	61.8	0.6
Pears	52.1	7.1	26.5	5.9	55.8	4.7
Tomatoes	62.8	23.7	16.3	59.7	—	0.3
Onions	90.2	28.8	40.5	19.7	7.6	3.4
Milk, preserved not dried	−18.8*	96.8	—	—	—	3.2
Cream, fresh	6.8	100.0	—	—	—	—
Milk, powder	30.1	23.9	15.9	—	59.7	0.5
Cream, preserved	43.2	99.3	—	—	—	0.7
Cheese, other than blue-vein	54.3	15.6	2.9	7.4	73.4	0.7
Butter	98.1†	31.0	11.5	—	56.6	0.9

* Net export, mainly to tropical and subtropical countries.
† All imports as percentage of all home production of all cheese types.
Zone 1. Belgium, Channel Islands, Denmark, Eire, Netherlands, Norway.
Zone 2. Austria, Finland, France, Italy, Poland, Spain, Sweden, Switzerland, Yugoslavia.
Zone 3. Algeria, Canada, Canary Islands, Cyprus, Egypt, Lebanon, Libya, Malta, Morocco, United States.
Zone 4. Argentina, Australia, Chile, East Africa, New Zealand, South Africa.
Sources: Official production and trade returns, with some supplementation from unofficial sources.
Reprinted with permission from *Rural Settlement and Land Use* (1st edition) by M. D. I. Chisholm (1962) published by Hutchinson, London, table 11.

LAND-USE RINGS DISTORTED

Von Thünen was aware of factors likely to distort his model agricultural landscape. Perhaps the best-known distortions are caused by the presence of a second market, differences in soil fertility, and a navigable river flowing through the area. Another factor which has recently been added to the long list is known as suboptimal behaviour. This occurs where some farmers produce crops and raise livestock which give less than the economic rent but, as far as the farmers are concerned, bring in a satisfactory income. It may arise because farmers lack information on what is the most economical enterprise in which to indulge, because they have not the funds to act on such information even if they were to get it, or because they do not wish to obtain the largest possible return from investments. The idea that man generally makes what he sees as satisfactory decisions, rather than what would in fact be the best decision from an economic point of view, was put forward by H. A. Simon in 1957 in his book *Models of Man* and has been taken up by geographers who describe it as suboptimal behaviour. A case in point is the study made by Julien Wolpert of farming in the middle region of Sweden in the early 1960s. Wolpert showed that very few farms achieve the highest possible productivity, as measured by the income from an hour's labour, the average farmer attaining some 60 per cent of the potential value (Figures 1.20a and b).

1a   Study Figures 1.21a, b, c, and d, which indicate how Von Thünen-type land-use rings may be distorted by patches of fertile and infertile soil, by competing markets, and by lines of cheaper transport such as navigable rivers and railways. Suggest how each of the complicating factors operates.

b   Map on Figure 1.21e the land-use pattern you think would arise when all three complicating factors – soil fertility, competing markets, and lines of cheaper transport – act together.

c   Describe how you think suboptimal behaviour by some of the farmers in the region might influence the land-use patterns.

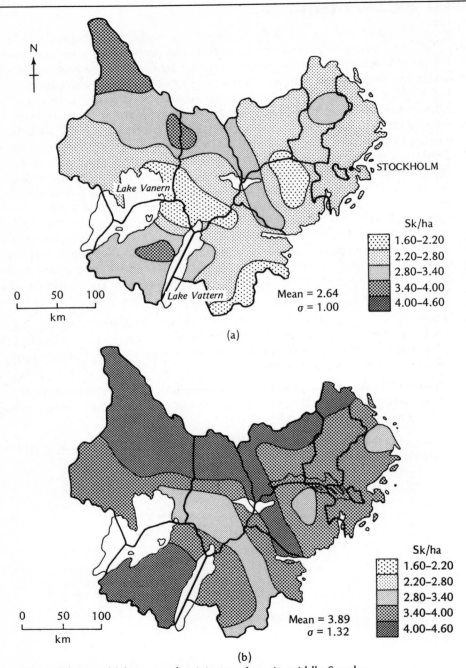

Figure 1.20 (a) Actual labour productivity per farm in middle Sweden
(b) Potential labour productivity per farm in middle Sweden
Reprinted with permission from the *Annals* of the Association of American Geographers, volume 54, 1964, J. Wolpert.

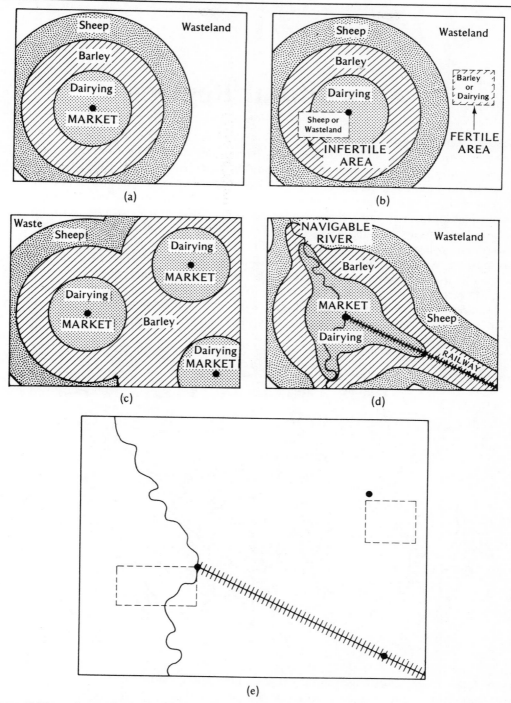

Figure 1.21   Some distortions of Von Thünen's rings of land use

Agricultural Regions

Agricultural patterns are fashioned by many economic, behavioural, and physical processes. Nonetheless, agricultural regions, each characterized by certain farming techniques and patterns of land use, can be identified. On a world scale, Whittlesey in 1936 used crop and livestock associations, the intensity of land use, the degree of processing and marketing of a product, the method of production, and the complex of farm structures associated with the farm enterprise, and came up with the broad categories of agriculture, such as nomadic herding, livestock ranching, subsistence crop and livestock ranching, commercial crop and livestock ranching, dairy farming, Mediterranean agriculture, and plantation agriculture. On a smaller scale, agricultural regions can be identified by looking at crop and livestock combinations. J. C. Weaver studied farming in America's Midwest in the 1950s and devised an enterprise classification scheme. He found that because crops are generally grown in combinations, regions like 'the Corn Belt' and 'the Cotton Belt' (Figure 1.22) are not monocultural.

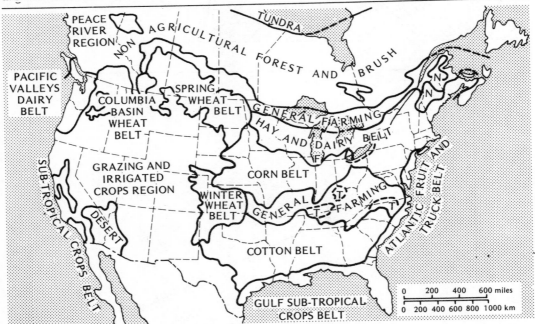

Figure 1.22   Agricultural regions in the United States and Canada.
From *North America: A Geography of Canada and the United States*, fourth edition, by J. H. Paterson. © 1960. 1962. 1965. 1970 by Oxford University Press. Reprinted by permission

ENTERPRISE COMBINATIONS IN ENGLAND AND WALES

J. T. Coppock adapted Weaver's technique to produce enterprise combination regions for England and Wales using data gathered in 1958 from 350 agricultural advisory districts; his results were published in 1964 as *An Agricultural Atlas of England and Wales*. Figure 1.23a portrays the leading enterprises in different parts of England and Wales as mapped by Coppock.

But this figure shows leading enterprises only. Figure 1.23b, on the other hand, maps the crops and livestock found in combination with the leading enterprises.

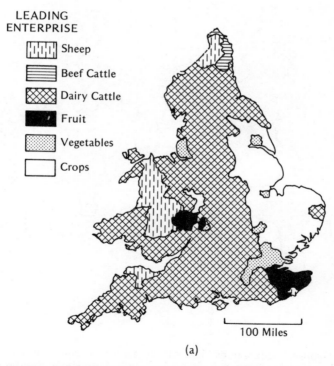

(a)

Figure 1.23a   Leading enterprises in England and Wales (seven enterprises)
Reprinted by permission of Faber and Faber Ltd from *An Agricultural Atlas of England and Wales* by J. T. Coppock, figure 200.

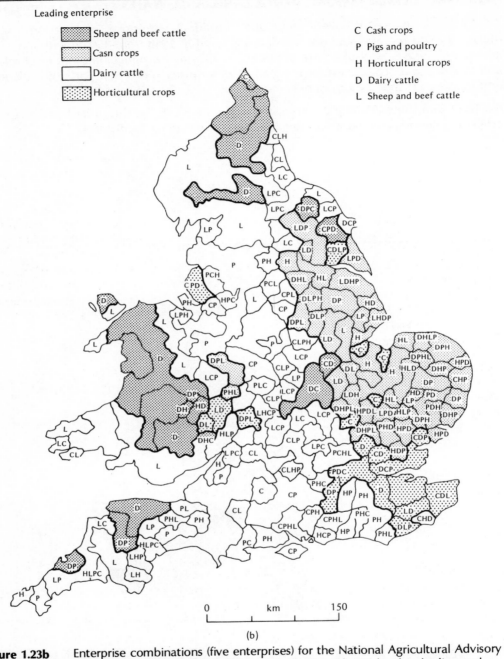

Leading enterprise

Sheep and beef cattle

Casn crops

Dairy cattle

Horticultural crops

C Cash crops
P Pigs and poultry
H Horticultural crops
D Dairy cattle
L Sheep and beef cattle

0   km   150

(b)

Figure 1.23b   Enterprise combinations (five enterprises) for the National Agricultural Advisory Districts of England and Wales, 1958. The leading enterprises are shown by the shading and the others in the combination systems by the overprinted letters.
Reprinted by permission of Faber and Faber Ltd. from *An Agricultural Atlas of England and Wales* by J. T. Coppock, figure 201.

The Coppock–Weaver technique entails classifying the land use on each farm. Data are collected on crop acreage and number of livestock (Table 1.8a).

Table 1.8a   Enterprise acreages on Beech Farm

Enterprise	Acreage
Wheat	93
Barley	240
Oil Seed	31
Forage peas	22
Herbage seed	63

The data are converted into units which measure the labour requirements per acre or per head expressed as standard man-days. Tables of standard labour requirements for various enterprises in England and Wales are published by the Ministry of Agriculture (for example, Table 1.9). On Beech Farm there are 93 acres of wheat, each acre of which, according to Table 1.9, requires two man-days a year; so in total the wheat grown on Beech Farm takes up 186 man-days' worth of labour. The standard labour requirements for other crops on Beech Farm are listed in Table 1.8b.

Table 1.8b

Enterprise	Standard man-days	Percentage
Wheat	186	18.4
Barley	480	47.2
Oil seed	93	9.2
Forage peas	66	6.5
Herbage seed	189	18.7
Totals	1014	100.0

The labour required for each crop is then expressed as a percentage of the total standard man-days for all enterprises on the farm (Table 1.8b). For instance, barley uses 480 out of 1014 standard man-days, or 47.2 per cent. The percentages are compared with a set of percentages derived from theoretical combinations of enterprises. For example, if Beech Farm were under one enterprise, obviously 100 per cent of the land would be given over to that enterprise; if Beech Farm were a two-leading-enterprise farm, it is assumed by the method that half the land would be given over to one of the leading enterprises and half the land to the other; where three leading enterprises are found, each is assumed to occupy a third of the farmland area; and so forth. These theoretical percentages are shown in the expected percentage row of Table 1.8c. The observed percentages in the same table are simply the data from Table 1.8b – for a two enterprise combination the percentages are 47.2 for barley and 18.7 for herbage seed.

Table 1.8c

	Monoculture	Two-crop combination		Three-crop combination			Four-crop combination				Five-crop combination				
Observed percentages	47.2	47.2	18.7	47.2	18.7	18.4	47.2	18.7	18.4	9.2	47.2	18.7	18.4	9.2	6.5
Expected percentages	100.0	50.0	50.0	33.3	33.3	33.3	25.0	25.0	25.0	25.0	20.0	20.0	20.0	20.0	20.0
Difference, d	−52.8	−2.8	−31.3	13.9	−14.6	−14.9	22.2	−6.3	−6.6	−15.8	27.2	−1.3	−1.6	−10.8	−13.5
Difference squared, d^2	2787.8	7.8	979.7	193.2	213.2	222.0	492.8	39.7	43.6	249.6	739.8	1.7	2.6	116.6	182.2
Sum of differences squared Σd^2	2787.8	987.5		628.4			825.7				1042.9				
$(\Sigma d^2)/n$	2787.8	493.8		209.5			206.4				208.6				

Table 1.9 Standard labour requirements in England and Wales, 1971

Crop	Standard man-days (per acre)	Livestock	Standard man-days (per head)
Wheat	2	Dairy cows in milk	10
Barley	2	Beef cows in milk	3
Oats	3	Bulls at A. I. centres	20
Potatoes	15	Ewes	0.7
Sugar beet	10	Breeding sows	4
Hops	70	Hens, geese, turkeys	0.1
Orchards grown commercially	23	Bees (per colony)	1
Strawberries	70		
Gooseberries	40		
Carrots	10		
Beetroot (red beet)	30		
Peas, green for market	30		
Peas, for processing	3		
Bulbs	100		
Crops under glass or sheds	1300		
Bare fallow	0.5		
Permanent grass	0.5		

Note: An additional 15 per cent should be added for essential maintenance and other indirect labour.

Source: Based on Ministry of Agriculture data.

To determine which of the theoretical combinations best matches the observed combinations, a statistical test, given by the formula

$$\frac{(\Sigma d^2)}{n}$$

is used where d is the difference between observed and theoretical percentages for each enterprise under varying enterprise combinations, and n is the number of enterprises in the combination – one, two, three, and so on. The value of $(\Sigma d^2)/n$ is worked out for each enterprise combination (Table 1.8c). The lowest $(\Sigma d^2)/n$ value is 206.4; this enables us to say that Beech Farm specializes in cash crops, the four leading enterprises being herbage seed, barley, wheat, and oil seed.

1 Examine Table 1.10, which is based on part of M. G. Day's sample study of farming in Hampshire (Figure 1.24). Classify the farm using the Coppock–Weaver method.

Table 1.10

Enterprise	Acreage
Wheat	93
Barley	29
Potatoes	30
Carrots	35
Total	187

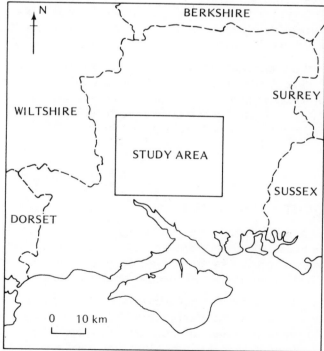

Figure 1.24   Location of the study area in Hampshire

Coppock classified the Hampshire study area as one in which dairy cattle was the leading enterprise in combination with cash crops, pigs and poultry, horticulture, sheep, and beef cattle. But this pattern may have changed between the date of Coppock's survey and the one made in 1977 because important land-use changes have taken place in Hampshire during the past fifteen years, with a general trend away from dairying towards the cultivation of grasses and other feed crops. Farms have also, by a process of reorganization, tended to become larger. In 1972, 80 per cent of the area was farmed by just 18 per cent of the farmers. Geologically, the region may be divided into the northeastern chalklands, and the southeastern Tertiary sands and gravels. The small farms which do still exist are clustered on the Tertiary beds around Southampton and especially in the New Forest. The largest and most prosperous farms are found on the chalk where one farmer may work 1000 acres. Over the past fifteen years, these large farms have turned to barley growing to such an extent that the region is now almost monocultural; this is partly because the water-storage and water-release properties of chalkland soils are well suited to barley. The barley crop is usually sold for malting, if it is not of high quality, or for stock feed. A general increase in demand for feedstuffs, and the effect of the European Community's farm pricing system, have increased the profitability of barley growing. But wheat is still grown in the region, and recently oil seed rape for margarine has been introduced. Off the chalkland, barley is usually found in association with dairying or the cultivation of herbage seed.

The trend towards rationalization of farming practices and increase in farm size is apparent in the region in the dairying enterprises which have declined in the face of better profits from cereals. Where carried out, dairying is usually a capital-intensive enterprise, with herring-bone parlours and labour-saving machines; it is common for one man to handle 80 to 100 cows.

So the land use in the Hampshire study area has changed to monocultural cereal growing on the chalklands with large, capital-intensive, well-organized farms. With this change, the use of fertilizers and chemical sprays, the economical use of labour, improved mechanization, cost-effective cultivation techniques, and a rationalization of marketing have become the order of the day.

FINDING THE RIGHT COMBINATION

A feature of British farming today is the willingness of farmers to experiment with new crops such as maize and especially with cash crops. The difficulty is to know how much land to give over to the new enterprises, bearing in mind the amount of labour needed to tend them and the likely cash returns from them.

Suppose a dairy farmer who has 100 hectares of land decides to plant some vegetables and some small fruit (raspberries and strawberries, for example). Assume he has the capital to convert the entire farm to these crops if he should wish. What is the hectareage of vegetables and of small fruit which will give him the greatest income? Table 1.11a summarizes the constraints the farmer must consider.

Table 1.11a   Labour constraints and net income

| Crop | Labour requirement (man-days/ha) | | Net income (£/ha) |
	Spring	Summer	
Small fruit	10	100	200
Vegetables	20	70	200
Number of man-days farmer has available	500	2000	

The problem is to maximize the income from small fruit and vegetables. We may write this as

(200 × small fruit) + (200 × vegetables) = maximum

where 200 × small fruit is the value of small fruit grown (£200 × hectares of small fruit) and 200 × vegetables is the value of vegetables grown (£200 × hectares of small vegetables). Two constraints to incorporate in the solution are the availability of labour and the amount of land available (the size of the farm). The labour constraints for spring and summer may be written (see Table 1.11a) as

(10 × small fruit) + (20 × vegetables) ≤ 500
(100 × small fruit) + (70 × vegetables) ≤ 2000

which means that the total time cultivating small fruit and vegetables in the spring must be equal to or less than 500 man-days and in the summer equal to or less than 2000 man-days. The land constraint (size of farm) is

hectares of small fruit + hectares of vegetables ≤ 100

The solution to the problem involves three steps.

Step one   Construct a graph with one axis as hectares of small fruit and the other axis as hectares of vegetables (Figure 1.25). Plot the constraints on this graph like this:

(a) Land constraint. If the farm were entirely given over to small fruit there would, obviously, be 100 hectares of small fruit and 0 hectares of vegetables; this fixes point A in Figure 1.25. Similarly, if the farm were all under vegetables, there would be 0 hectares of small fruit and 100 hectares of vegetables; this fixes point B on the graph. Join points A and B by a straight line; this line represents all possible combinations of small fruit and vegetables within 100 hectares.

(b) Labour constraints. In spring, 500 man-days could be devoted to small fruit cultivation in which case 500/10 = 50 hectares of small fruit could be grown; this fixes point C on the graph. On the other hand, 500 man-days could be spent growing vegetables and 500/20 = 25 hectares of vegetables could be planted; this fixes point D on the graph. Join points C and D by a straight line. The line so formed is the spring labour constraint and represents all possible combinations of small fruit and vegetables which could be grown utilizing fully the available spring labour.

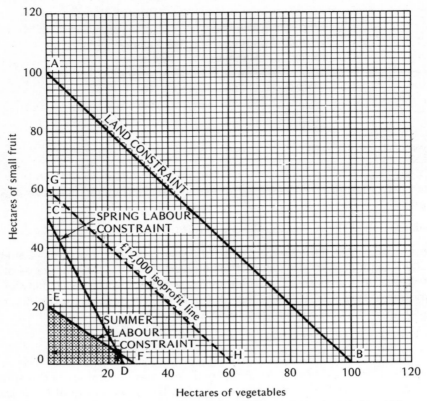

Figure 1.25   Constraints and profits for the growing of small fruit and vegetables

In summer, 2000 man-days could be devoted to small fruit production, so 2000/100 = 20 hectares of small fruit could be grown; this fixes point E on the graph. At the other extreme, with all labour used for vegetable cultivation, 2000/70 = 28.6 hectares of vegetables could be grown; this fixes point F on the graph. Join points E and F by a straight line. This line is the summer labour constraint and represents all possible combinations of small fruit and vegetables which could be grown utilizing fully the available summer labour.

Step two   Identify on the graph the zone of feasible production. This is the shaded area which lies beneath all three constraint lines.

Step three   Find the point in the feasible production zone which gives the largest net income. To do this, isoprofit lines must be drawn in. Take for example the £12,000 isoprofit line. This is plotted in two steps. First of all, work out the number of hectares of small fruit required to make £12,000. Small fruit yields £200/ha so to make £12,000 would take 60 hectares of small fruit; this fixes point G on the graph.

Secondly, work out the number of hectares of vegetables required to make £12,000. Vegetables yield £200/ha so to make £12,000 would take 60 hectares of vegetables; this fixes point H on the graph. Join points G and H to produce the £12,000 isoprofit line which shows all combinations of small fruit and vegetables which would give a net income of £12,000. All other isoprofit lines on the graph will lie parallel to this one. Find the isoprofit line with the largest value in the zone of feasible production; it is roughly £5400. Read off the hectares of each crop which will give this income: 4 hectares of small fruit and 23 hectares of vegetables. Notice that this optimum solution, being very constrained by labour supply, by no means uses all the available land.

1 The farmer decides to let people pick their own fruit in summer. This frees money to take on additional labour in spring and allows more farmhands to work with vegetables in summer. The new constraints are shown in Table 1.11b.

Table 1.11b   New constraints with pick-your-own fruit

Crop	Labour requirement (man-days/ha)		Net income (£/ha)
	Spring	Summer	
Small fruit	10	100	200
Vegetables	20	70	200
Number of man-days farmer has available	900	4000	

What is the optimum area to devote to each crop now and how much net income does the farmer stand to make?

2 A neighbouring farmer has the same labour supply as in Table 1.11b but he has just 40 hectares of land. What difference does this make?

It can be seen that regional 'belts' of agriculture may be identified in which one enterprise is dominant. But these belts are not necessarily areas of monoculture; other enterprises are commonly carried out in combination with the leading enterprise. The question of which combination of enterprises will yield the biggest profit for a given set of physical and economic constraints may be tackled by a mathematical technique called linear programming – this is what you used in the small fruit and vegetables problem.

Agricultural Change

Agricultural patterns seldom remain unchanged. They are from time to time refashioned by agricultural innovations – new strains of crops and livestock, new fertilizers, and new machinery; by the economic climate of demand; and even by the whims and fancies of farmers themselves.

THE SPREAD OF AGRICULTURAL PRACTICES

In 1952, in a now classic study, the Swedish geographer Torsten Hägerstrand examined how a number of agricultural innovations, including the adoption of farm subsidies and tuberculosis inoculations in cattle, spread or diffuse through a region. Take the adoption of farm subsidies in the province of Östergötland in Sweden. The Swedish government first introduced the subsidy in 1928 to encourage farmers with small farms to improve and enclose their pastures and to discourage farmers from allowing their cattle to forage in open woodland during summer months. The distribution of farmers who had adopted the subsidy by 1929 is shown in Figure 1.26a and for the following three years in Figures 1.26b, c, and d. (The figures are gridded into 5 × 5 kilometre cells, that is, 25 square kilometres, and the numbers indicate the total number of farms that have accepted the subsidy, accumulated from one date to the next.)

1 What impact do you think the subsidy would have had on the amount and distribution of open woodland in Östergötland?

2 The first subsidies were adopted in the west-central part of the area. To show the nature of the spread from this point, either (a) on a copy of Figure 1.26, draw isolines of adopters for each year. Use these contour intervals: less than 1, 4.5, and 9.5. Or (b) draw transects showing number of adopters from cells A12 to L1 using a different colour for each year (Figure 1.27). And (c) for either (a) or (b), discuss the nature of the spread.

Hägerstrand found that the clustering of adopters around the area where the farm subsidy was first adopted reflected the way in which information was spread from one farmer to another: information was spread by personal contact, so a farmer who knew about the subsidy would be more likely to tell a farmer on a neighbouring farm of it than a farmer living some distance away; this is called the neighbourhood effect (Figure 1.28). To simulate the spread of information, Hägerstrand used a Monte Carlo technique (see *Settlements*, chapter one, Harper & Row, 1979). Figure 1.29 shows Hägerstrand's simulation for 1932 and the closeness of fit with the actual pattern of adoption (Figure 1.26d) is high.

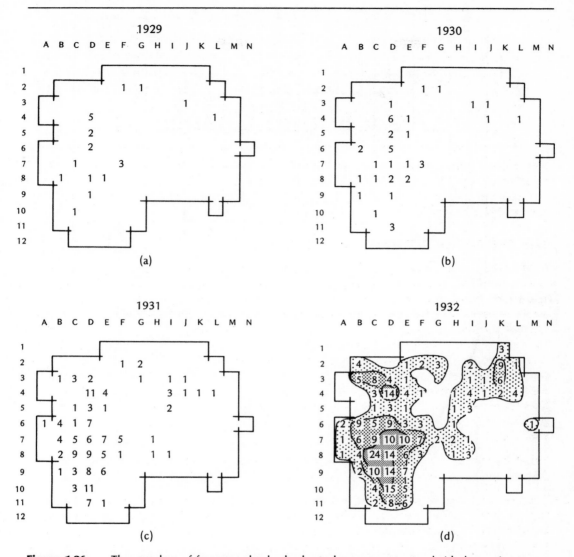

Figure 1.26 The number of farmers who had adopted a government subsidy by a given date in the Östergötland region of Sweden
Reprinted from *Northwestern University Studies in Geography*, number 13, by permission of Northwestern University Press.

In recent years, research has cast doubts on the validity of some of Hägerstrand's assumptions: Hägerstrand's studies were of rural society prior to the growth of modern communication networks and information sources, and ignored the willingness of individual farmers to seek information in farming magazines and so on. Tornqvist in 1968 argued that mass media are responsible for an adopter's initial awareness of an innovation, whereas personal contacts, which Hägerstrand thought were of prime

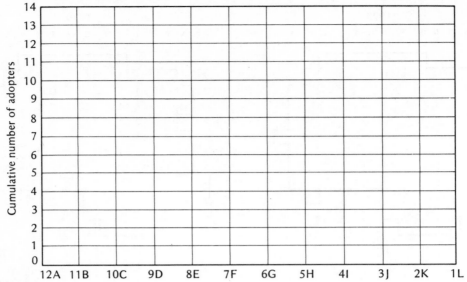

Figure 1.27

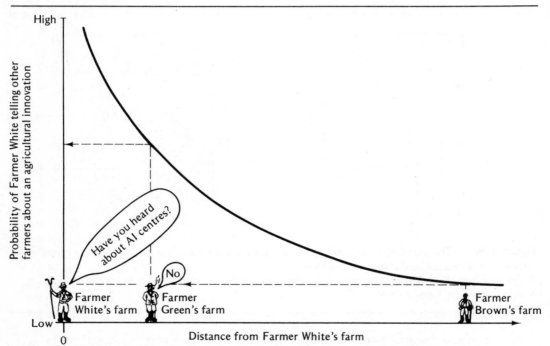

Figure 1.28 A curve showing how the probability of one farmer telling another about an agricultural innovation – artificial insemination centres for instance – falls with distance from the farm of the farmer doing the telling. This is called the neighbourhood effect. In the example shown, Farmer White is more likely to tell his close neighbours like Farmer Green, about artificial insemination centres than he is his more distant neighbours, like Farmer Brown.

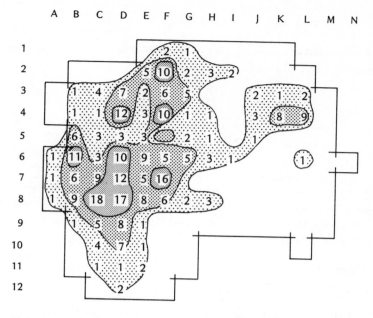

Figure 1.29 The number of farmers who had adopted the government subsidy by 1932 as predicted by a run of Hägerstrand's Monte Carlo model.
Reprinted from *Northwestern University Studies in Geography*, number 13, by permission of Northwestern University Press.

importance, would influence the farmer in weighing up whether or not to adopt. Allan Findlay and Duncan Maclennan examined the findings of two agricultural surveys conducted in Scotland. The Kintyre survey looked into the spread of bulk milk tanks and mechanical byre muckers during the period 1962 to 1971; the Ayrshire survey looked into the spread of barley fodder, milking pipelines, and brucellosis eradication during the period 1963 to 1975. The neighbourhood effect was not always important in explaining the spread of the agricultural innovations because personal mobility was high; and in Kintyre, contact with relatives was more important than contact with neighbours. In Ayrshire, farmers believed brucellosis was more difficult to eradicate in larger herds and this explains why the larger herds had later dates of brucellosis eradication. Economics also played a part in the patterns of adoption. For example, in Kintyre, the cost of bulk tanks reduced their adoption on small farms; but the spread of milk tanks hastened the adoption of pipeline milking systems, which shows how the pattern of adoption of one innovation may follow the pattern of adoption of an earlier innovation.

3 Examine Tables 1.12a and 1.12b and assess the validity of Tornqvist's and Hägerstrand's assumptions about the nature of information adoption for the Kintyre and Ayrshire survey results.

56

Table 1.12a Initial information sources, Kintyre

Source	Bulk tank %	Byre mucker %
Agricultural advisers	22.4	2.6
Salesmen	12.1	36.8
Press	56.9	15.8
Mass media	0	0
Neighbours/relatives	8.6	42.1
Others	0	2.6
Total	100.0	99.9

Reprinted with permission from 'Innovation diffusion at the micro-scale: a reconsideration of information and economic factors' by A. Findlay and D. Maclennan (1978). *Area,* 10(4), 309–314, table 1. Institute of British Geographers.

Table 1.12b Information sources during evaluation of an innovation, Kintyre and Ayrshire

	Bulk tanks %	Byre muckers %	Pipelines %	Brucellosis eradication %	Barley fodder %
Agricultural advisers	28.3	8.8	36.8	36.4	9.7
Salesmen	43.4	48.8	8.0	0.0	0.0
Press	11.3	7.0	13.0	5.5	9.7
Mass media	0.0	2.3	0.0	5.5	12.2
Neighbours	17.0	32.5	39.5	5.5	48.7
Vet	0.0	0.0	0.0	47.3	2.4
Others	0.0	0.0	2.5	0.0	17.1

Reprinted with permission, from 'Innovation diffusion at the micro-scale: a reconsideration of information and economic factors' by A. Findlay and D. Maclennan (1978). *Area,* 10(4), 309–314, table 2.

4 In 1960, Svi Griliches examined the adoption of hybrid maize seed by farmers in the United States. Each of the maize-producing states adopted hybrid maize at different dates and the rate at which hybrid maize spread from one farm to another within a state varied.

To illustrate the progressive adoption of hybrid maize during the period 1934 to 1958 in five selected states (Figure 1.30a), construct five cumulative frequency curves on Figure 1.30b by plotting the data from Table 1.13 and joining the points for each state. The average for the United States has already been plotted.

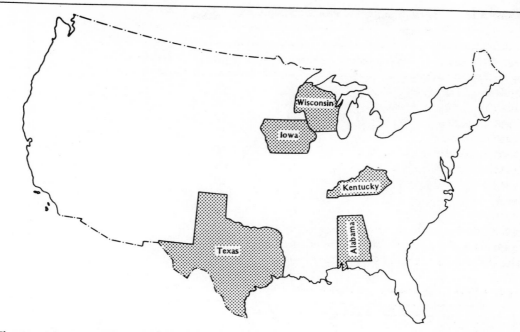

Figure 1.30a   Location of the five selected states

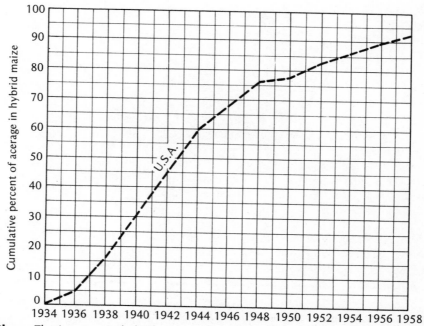

Figure 1.30b   The increase in hybrid maize acreage in five selected states and in the whole USA (already plotted)

58

Table 1.13 Acreage of hybrid maize as percentage of total maize acreage

Year	Iowa	Wisconsin	Kentucky	Texas	Alabama
1934	4	2	—	—	—
1936	14	7	—	—	—
1938	50	23	4	—	—
1940	88	55	8	—	—
1942	97	76	21	2	—
1944	99	85	48	4	3
1946		92	71	21	5
1948		94	82	52	10
1950		95	87	57	18
1952		96	89	70	27
1954		97	91	72	58
1956		98	94	78	72
1958		99	96	85	84

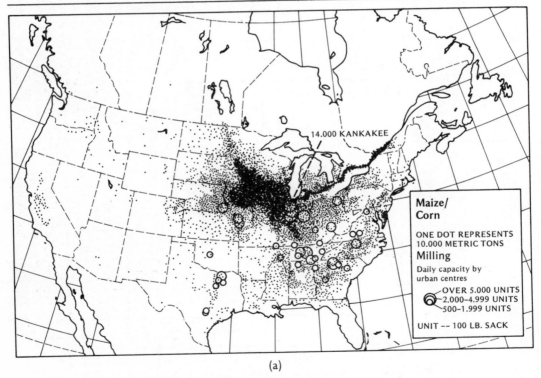

(a)

Figure 1.31a Maize-producing areas in the United States and Canada
Reprinted with permission from *Oxford Regional Economic Atlas: United States and Canada* published by Oxford University Press, London, p. 78.

Griliches attributed differences in dates of initial adoption of hybrid maize to supply factors, arguing that as a special brand of hybrid maize seed has to be developed to suit each environment within the United States, some regions received the seed before others. The seed producers, motivated by profit, developed hybrid maize seed for the chief maize-growing areas first (Figures 1.31a and b). Within each state, the rate of adoption on individual farms reflected other factors: in Iowa, where the soils are generally fertile, the new maize produced much higher yields than the old variety – a change-over could be made with profit; in Texas, the generally poor soils meant that the difference in yield between new and old varieties of maize was not so big as in Iowa and farmers were more reluctant about making a change-over.

Figure 1.31b   Maize production in Illinois. Reproduced with permission from Aerofilms

The effects of new crops, fertilizers, and equipment on land-use patterns can be difficult to assess. Consider a hypothetical case. Two farms, one belonging to Mr. Gumby and the other to Mr. Welly, both 100 hectares in size and both with all 100 hectares under cultivation, lie in a region in which the economic rents for wheat, oats, and maize are as listed in Table 1.14a.

Table 1.14a Economic rents

Crop	Gumby's farm (£/ha)	Welly's farm (£/ha)
Wheat	60	60
Oats	45	45
Maize	30	50

5 Assuming both farmers wish to make as much money as possible, what will be the land use under the conditions shown?

A new maize fertilizer is developed at a cost of £1.50 a bag. Table 1.14b indicates the increase in the yield of maize per hectare over the current yield. When the fertilizer is available to both farmers, will the land use change? To answer this

Table 1.14b

Number of bags of fertilizer (per hectare)	Increase in maize yield (tonnes per hectare)	
	Gumby's farm	Welly's farm
0	0	0
1	0.04	0.08
2	0.12	0.24
3	0.25	0.40
4	0.27	0.60
5	0.30	0.80
6	0.34	1.20
7	0.35	0.80
8	0.36	0.77
9	0.36	0.76

question the following method may be adopted. The first step is to calculate the change in the value of the output for the addition of each extra bag of fertilizer. Taking Gumby's farm first, the addition of one bag of fertilizer produces an extra yield of 0.04 tonnes of maize. Assuming maize fetches £20 a tonne at market, this means an extra £0.80 gross for farmer Gumby. On adding another bag of fertilizer on top of the one he has already added, Mr. Gumby's maize yield increases to 0.12 tonnes; this is 0.08 tonnes more than the increase from the first bag of fertilizer and grosses £1.60. On Welly's farm, investment in one bag of fertilizer increases the yield by 0.08 tonnes which brings in an extra £1.60 gross a hectare. On Welly's farm, a second bag of fertilizer brings in an extra £3.20; this is calculated as $(0.24 - 0.08) \times £20$. These values are called the value of the marginal product.

6 Evaluate the marginal product for each additional bag of fertilizer on Welly's farm, putting the results in Table 1.14c.

Table 1.14c

Quantity of fertilizer (bags)	Value of the marginal product (£/ha)	
	Gumby's farm	Welly's farm
1	0.80	1.60
2	1.60	3.20
3	2.60	3.20
4	0.40	
5	0.60	
6	0.80	
7	0.20	
8	0.20	
9	0.00	

On Gumby's farm, the optimum application of maize fertilizer is three bags for every hectare, the addition of more bags yielding less value of output than the cost of the bag of fertilizer.

The increase in economic rent at optimum fertilizer application levels is found by adding to the current economic rent for maize, £30 on Gumby's farm (Table 1.14a), the value of the increased production (this can be worked out from Table 1.14b), and subtracting from it the cost of fertilizer:

Gumby's farm

Value of increased maize yield (0.25 × £20)	£5.00
Cost of fertilizer (£1.50 × 3)	£4.50
Net increase in economic rent	£0.50
Total economic rent from maize-growing with fertilizer (£30 + £0.50)	£30.50

7   Find the total economic rent from maize growing at optimum fertilizer application on Welly's farm. What land-use changes would likely follow if the fertilizer became available to the farmers?

SUPPLY, DEMAND, AND AGRICULTURAL CHANGE

Wild fluctuations in prices for agricultural products can play havoc with the agricultural industry, so many governments try to keep prices steady. Consider the hypothetical supply-and-demand schedules for wheat depicted in Figure 1.32. At the outset, the price of wheat is £50 per tonne and the demand is for a thousand tonnes, all of which is met by farm production. A sudden increase in demand, to fifteen hundred tonnes, cannot immediately be met by the farmers – land must be ploughed and seeded, and the crop given time to grow and then be harvested. The supply is said to be inelastic. The result is an increase in the price of wheat (Figure 1.32). The higher price for wheat encourages farmers to grow more of it. In

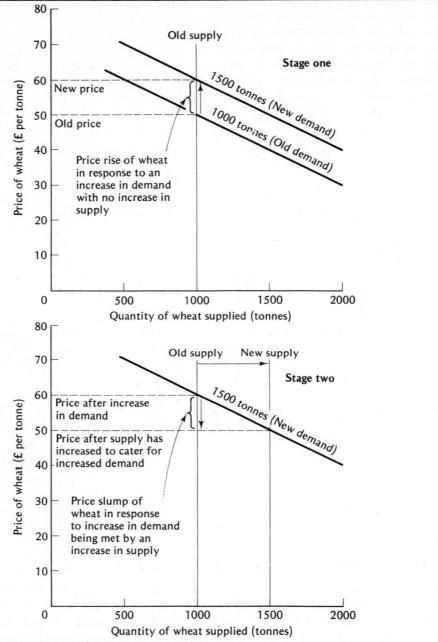

Figure 1.32 Hypothetical demand-and-supply schedules for wheat. The vertical axes show the price of a tonne of wheat. The horizontal axes show the total quantity of wheat supplied by farmers. The oblique lines show the old demand and the new demand for wheat. The following relationship between supply and demand is described by the demand curves. When supply just meets demand, the price of wheat is £50 a tonne. When supply falls short of demand, the price of wheat is more than £50 a tonne. The larger the wheat shortage, the higher the price of wheat. When supply exceeds demand, the price of wheat is less than £50 a tonne. The greater the surplus of wheat, the lower the price of wheat.

geographical terms an expansion of the area given over to wheat, or a more intensive use of the existing wheat land, can be expected. Figure 1.32 shows that, assuming the new demand for wheat remains fixed, the price will then fall by an amount which depends on the steepness (elasticity) of the demand curve. The reaction of farmers will be to cut back on the area of wheat production and grow other crops instead, or to grow wheat less intensively. In some cases, especially where farmers do not have much freedom of choice, farmers may actually increase the amount of land under the crop whose price is falling to earn enough money to stay in business.

Clearly then, price fluctuations can be harmful. The British government, prior to British entry into the Common Market, operated a system of deficiency payments. If the world price of a commodity fell, then the difference between the world price and a guaranteed price, decided upon by the Ministry of Agriculture and farmers as being fair, was made up by the government. In this way the British agricultural industry was given more stability than it otherwise might have had.

1 Examine Figure 1.33 and attempt to explain the changes in the amount of land under different enterprises since 1866.

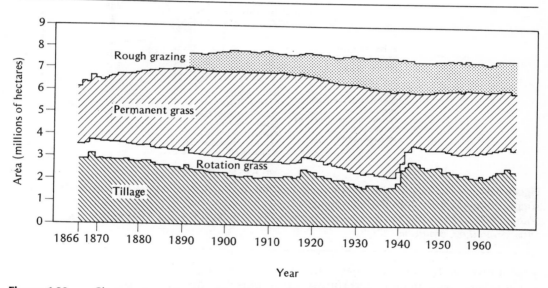

Figure 1.33   Changes in agricultural land use in England and Wales since 1866
Reprinted with permission from *The British Isles* (5th edition) by G. H. Dury (1973) published by Heinemann, London, figure 5.2.

The European Community operates a price guarantee system, the aim of which is to support the policy of increasing agricultural production, ensuring a reasonable standard of living for farmers, and keeping prices acceptable to consumers.

2 Carefully study Figure 1.34 and then explain how you think the cash fed into agriculture in the Common Market may alter existing patterns of land use. Illustrate

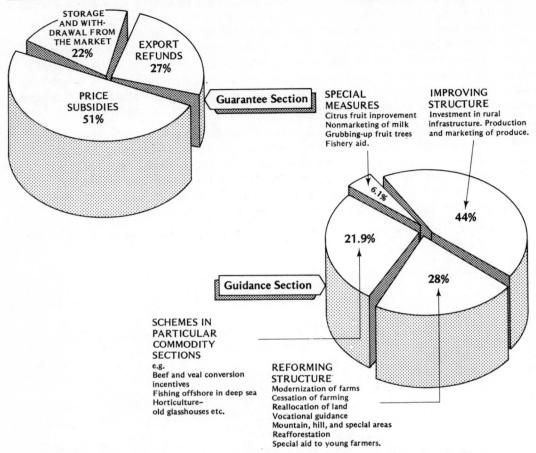

Figure 1.34 The cost of the common agricultural policy. Annual community expenditure in support of agriculture from the farm fund.
Source: Official Journal of the European Communities Vol. 20. 1979 and *The Agricultural Situation in the Community*, 1978.

your answer with examples. Bear in mind that the price guarantee system has occasionally "hiccoughed" and thrown up surpluses of milk and butter.

In Canada, the federal government has attempted to alleviate serious economic problems in depressed rural areas. Farms which bring in low incomes are widespread in Canada but they are particularly numerous where soil or climate make farming difficult. In the limestone plains of southern Ontario, where the soils are shallow and stony, and suffer from a shortage of water in the summer, crop yields are frequently low. In the Dummer Moraine region, the presence of vast expanses of boulders adds to the problems of farming.

3a With reference to Table 1.15 construct a choropleth map (Figure 1.35b) to show the concentration of low-income farms in southern Ontario.

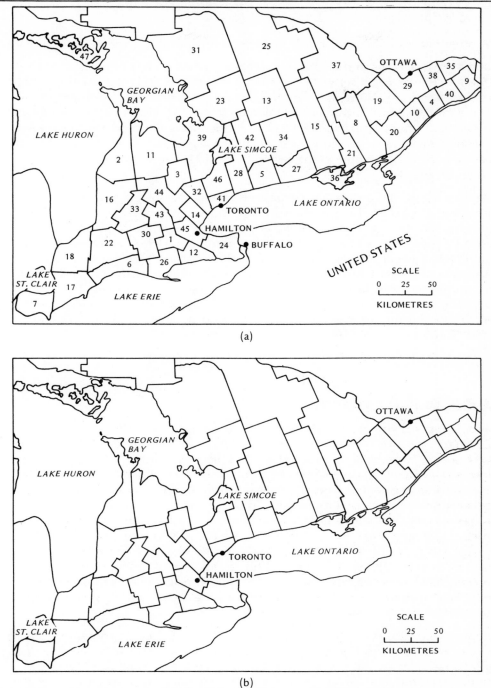

(a)

(b)

Figure 1.35 From *Poverty Pockets* by R. C. Langman, reprinted with permission of the Canadian publishers, McClelland and Stewart Ltd, Toronto

3b To what extent are areas with low-income farms associated with outcrops of limestone (Figure 1.36)?

c What other factors could explain the location of low-income farms?

A number of townships in the limestone region reached a peak population in the nineteenth century when settlers were eager to own land. But they found it difficult to make their farms pay. Nowadays, half the farms in the limestone region are run by part-time farmers and it is this lack of full-time farming which partly explains why a large number of agricultural workers have left the region. Furthermore, it is the young that tend to move to neighbouring industrial cities such as Toronto. This has led to a preponderance of old people in the region (Table 1.15) and a rundown in the provision of services.

In the early 1960s, the problem of depressed farming regions like that in southern Ontario was tackled by the Canadian government which passed an Agricultural and Rural Development Act. This act aimed at raising low incomes in the agricultural industry by promoting a more efficient use of resources. Unfortunately, in the southern Ontario region its effects were small because the measures taken, namely the introduction of modern farm equipment and the combining of the originally small farms into bigger units, failed to create jobs. Some measure of success has been achieved by the Area Development Agency which was set up in 1963 to provide incentives for industry to locate in areas of high unemployment. By 1971 it had created 17,000 jobs in Ontario, but the region remains a poor relation.

CHANGES IN FARM STRUCTURE

Modern farmers in western Europe have inherited small farms. The average farm size is 12 hectares but two-thirds of the farms are less than 10 hectares in size. In many parts of Britain, farms tend to be larger than in the rest of western Europe because enclosure movements in the eighteenth and nineteenth centuries swept away the open fields and fragmented plots of medieval farmers and replaced them with large, rectangular fields surrounded by hedges. To run efficiently in the present day, old fields, farms, and farmhouses need restructuring. Many governments have encouraged restructuring programmes. But two problems beset plans to restructure farms. In the first place, the farms in western Europe being on the small side, they barely bring in enough income to support a farmer and his family. (An exception to this is intensively cultivated, small market garden plots.) In the second place, many farms consist of strips of land scattered over a large area and so are time-consuming and uneconomic to run, and do not lend themselves to mechanization.

Attempts to increase the size of farms in some western European countries includes legislation which ensures that when farmland falls vacant it should be used to enlarge neighbouring holdings. But this is a slow process and the average size of farms in western Europe has increased at a mere hectare each decade. Agencies also operate which buy up land as it comes onto the market, hold onto it for a while so as to improve it with things like new roads and buildings, then sell it off for the enlargement

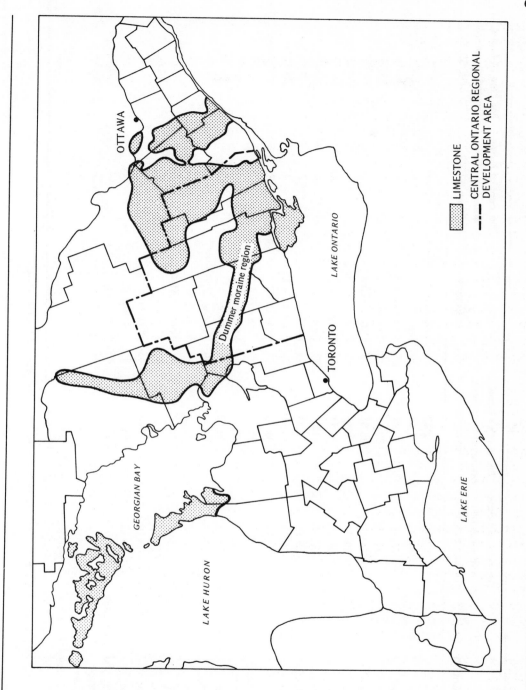

Figure 1.36 Limestone areas in Ontario
Adapted from *Poverty Pockets,* by R. C. Langman, published by McClelland and Stewart (1975).

Table 1.15

	County	Per cent of farms with sales less than $2500	Per cent of farms operated by people over 55		County	Per cent of farms with sales less than $2500	Per cent of farms operated by people over 55
1	Brant	25.68	32.42	25	Nipissing	51.08	30.51
2	Bruce	19.36	34.47	26	Norfolk	19.33	27.03
3	Dufferin	27.38	32.91	27	Northumberland	35.52	38.81
4	Dundas	24.45	38.66	28	Ontario	28.22	33.34
5	Durham	38.04	35.51	29	Ottawa-Carleton	35.57	36.79
6	Elgin	20.77	30.55	30	Oxford	15.62	30.92
7	Essex	26.25	35.11	31	Parry Sound	59.42	36.92
8	Frontenac	50.21	41.71	32	Peel	35.66	33.39
9	Glengarry	28.55	38.31	33	Perth	14.29	28.93
10	Grenville	51.99	41.16	34	Peterborough	44.47	36.15
11	Grey	29.53	36.43	35	Prescott	24.13	29.55
12	Haldimand	34.61	35.50	36	Prince Edward	33.30	36.98
13	Haliburton	77.88	45.19	37	Renfrew	49.44	40.56
14	Halton	38.89	34.02	38	Russell	24.54	23.10
15	Hastings	47.67	38.93	39	Simcoe	35.21	34.97
16	Huron	15.42	31.37	40	Stormont	35.54	36.84
17	Kent	13.63	34.07	41	Toronto Metro. Munic.	43.51	52.77
18	Lambton	25.42	34.51	42	Victoria	40.08	40.68
19	Lanark	43.19	41.61	43	Waterloo	18.97	17.20
20	Leeds	40.12	43.52	44	Wellington	24.72	31.78
21	Lennox and Addington	47.82	41.19	45	Wentworth	38.43	35.42
22	Middlesex	23.12	35.55	46	York	40.82	31.74
23	Muskoka	60.59	49.26				
24	Niagara Reg. Munic.	38.40	35.36				

Source: Census of Canada, Catalogues 96–722 (September 1972) and 96–723 (October 1972).

of surrounding farms. In France, the *Sociétés d'Aménagement Foncier et d'Établissement* (SAFER) is such an agency and acts to ensure that average farm size is on the up and up.

Attempts to ensure the amalgamation of scattered plots into compact holdings have met with a good degree of success. A variety of schemes of plot consolidation, ranging from the simple swapping of parcels of land to ambitious programmes of rural management, exist in all countries of western Europe. In France, some piecemeal plot consolidation took place after 1918 but the main thrust came in 1941 when an official policy of plot consolidation (*remembrement*) was started. Forty per cent, or some 14 million hectares, of French farmland was considered to be in need of reorganization. By 1969, over 6 million hectares had been consolidated. An example of consolidation is shown in Figure 1.37, though, because newly consolidated plots must be of the same productive value as the strips formerly held by each landowner, not many schemes have been as successful as this one.

Brittany: A Case Study

Farms in Brittany are small affairs. The average size reported by Louis Smith in 1969 was 11.8 hectares. In the *Pays de Redon*, several holdings were less than 6 hectares in size, each being divided into as many as twenty-eight scattered plots, many with individual strips of land measuring less than 3-metres across. Government-financed plot consolidation did not operate in Brittany until 1952. By 1972, 30 per cent of the farmland that needed it had been consolidated. A local SAFER agency has also been set up in Brittany to assist in the restructuring and enlarging of farms. During the period from 1962 to 1968, the activities of SAFER increased the size of farms by 36 per cent. These structural changes in agriculture in Brittany have led to the large-scale disappearance of *bocage* country – scattered woodland and irregularly shaped fields separated by hedgerows – much to the consternation of conservationists who question the merits of unbounded open land in an area of undulating terrain.

1 To what extent do you think structural changes in Breton agriculture have assisted the recent revival of the Breton economy? (See reference list at the end of this chapter.)

As well as the forms of structural change in agriculture in western Europe, there have been movements to change land ownership through land reform in Italy, Communist countries of eastern Europe, and countries such as Bolivia.

Land Reform In Italy

The aims of land reform in Italy were social and political in nature. They involved the breaking up of massive, underused estates to provide farms for landless agricultural workers. The result has been the reverse of farm enlargement and a headache for the Italian agricultural planners. Four-fifths of the land involved in reform was in the south of Italy. Some 700,000 hectares of land were redistributed among some 100,000 families between 1951 and 1962. The average size of the new farm was 7 hectares but varied between 4 hectares on irrigated land to 20 hectares on poor hill land.

The success of land reform in Italy varies from one region to another. Irrigated farms in the Metaponto region are successful. But in Calabria, many people considered living in isolated farmhouses to be lonely and unsafe. D. McEntire has described the

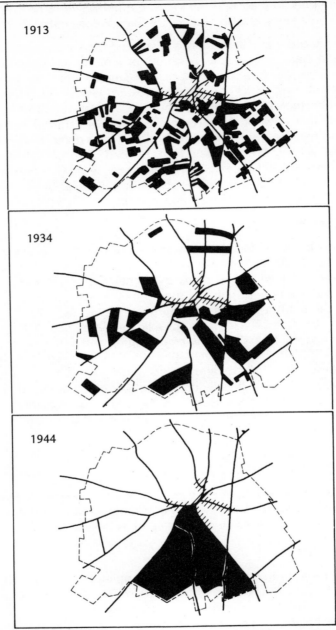

Figure 1.37 *Remembrement* of Le Bosquet *commune*, Somme *département*, northern France. Two phases of plot consolidation have taken place. In the first, each landowner still holds a number of blocks of land; this is usually as far as consolidation goes. In the second phase, large, unbroken blocks of land have been created.
Reprinted with permission from *Rural Geography* by H. D. Clout (1972) published by Pergamon Press Ltd., Oxford, figure 7.1.

conditions he found: 'Here [in Calabria] was a poor, unsanitary village, probably no better in its sanitary condition than 20 years before, but crowded with people, while the adjoining countryside was dotted with modern, commodious and empty houses.'

Land Reform In Bolivia

In 1950, 70 per cent of all agricultural holdings in Bolivia were less than 10 hectares in size but occupied a mere 0.4 per cent of the cultivated area. On the other hand, just 8 per cent of the farms were bigger than 500 hectares but occupied 95 per cent of the cultivated area. Land reform started in 1953 and resulted in the breaking up of the vast estates which had been run on feudal lines. Some medium-sized properties were also broken up but not all of them were. For instance, in the cold desert of the Altiplano, large farms were not broken into units of less than 350 hectares. By 1970, the land-reform programme had benefitted more than a quarter of a million farmers.

Notice that land-reform programmes seek social justice by taking power, property, and status from a 'select' social group and giving it to the 'workers'. In this sense, as John Galbraith the American economist has noted, land reform is revolutionary.

It can be seen that the pattern of agricultural land use may change, chiefly in response to an unpredictable economic climate. Some governments try to buffer adverse changes, which might be brought about by, say, a slump in the market price, by awarding subsidies. Small scale changes in agriculture may be brought about by the whims of farmers. The adoption of an innovation in agricultural technology, such as the introduction of artificial insemination in pigs of the growing of maize in England, tends to spread slowly from one farm to another. More drastic changes, involving the alteration of farm structure and land reform, are induced by political forces.

Further Reading

Geography: A Modern Synthesis, P. Haggett, Harper & Row (1979), Chapter 17.
Location in Space, P. Lloyd and P. Dicken, Harper & Row (1977), Chapters 2 and 3.
Spatial Analysis: Area Patterns, J. Blunden et al., Open University (1977), Unit 15.
Agricultural Geography, W. Morgan and R. J. C. Munton, Methuen (1971), Chapters 5, 7, and 8.
Rural Settlement and Land Use, M. D. Chisholm, Hutchinson (1979).
Regional Development in Western Europe, H. Clout (ed.), Wiley (1975), Chapters 4, 5, and 6.
Breton Farmyard Politics, M. Philiponneau, Geographical Magazine, (February 1975).

CHAPTER TWO
THE PHYSICAL GEOGRAPHY OF AGRICULTURE

Crops and farm animals, like their wild counterparts, occur in a vast array of habitats, each with its own combination of light intensity, soil moisture, temperature, nutrient levels, and so forth. Any one crop or beast is able to survive in a certain range of habitats. For every environmental factor there is a lower limit below which the species cannot grow at all, an optimum level at which growth is best, and a maximum level above which no growth occurs. Any species makes good between the upper and lower limits which define the ecological tolerance of that species to a particular environmental factor. An environmental factor which retards the growth of an organism is called a limiting factor, as first suggested by Justus von Leibig, an agricultural chemist who observed that the growth of a crop is hampered by whatever nutrient happens to be in short supply. A field of wheat may have abundant available phosphorus to give a high yield but may yield poorly because another nutrient, say nitrogen, is in short supply. No matter how much additional phosphorus is added in fertilizer, the yield will not increase. Only by adding nitrogen can the yield be increased, and it will do so in proportion to the amount of nitrogen added, up to an optimum level. By removing limiting factors one by one in the American Midwest, corn yields have been greatly improved. There is an upper acceptable limit to the amounts of nutrients in a field – the results of adding too much nitrogen in fertilizer could be disastrous. Clearly then, the growth of a crop (or farm animal) depends on a complex of interacting environmental factors. Should any one of these factors fall short of a lower limit or exceed an upper limit specific to the organisms concerned, then this factor is a limiting factor.

The limiting factors we shall consider in relation to agriculture are climate, soil, and biology.

Climatic Limitations

Sunlight and warmth are needed for crops to germinate, grow, and ripen. Grass needs a minimum temperature of 6°C before it will start to grow. Wheat, barley, and oats require a minimum temperature of 4°C. Rye will start to grow at temperatures as low as 2°C, but maize will not start growing unless the minimum temperature of 9°C is reached. The time taken for crops to attain maturity also varies from one crop to another. Barley, for instance, takes less time to mature than wheat and so may be cultivated in areas with a growing season too short for wheat to give good yields. The growing season is not simply the time between sowing and harvesting. A rough and ready guide to the length of the growing season is the number of days in a year when the minimum temperature needed for a particular crop to grow is exceeded. A better measure is provided by accumulated temperatures – the cumulative amount by which each day's temperature goes over the minimum temperature throughout the growing season. Wheat requires at least 1300 day-degrees of accumulated temperature. Frosts can destroy crops in spring and autumn. Grain crops in Finland, which is at the northern margin of grain cultivation, can be devastated by frosts one year in every forty. At a more local scale, the formation of frost pockets in valley bottoms can severely damage crops, fruit crops being exceptionally prone to frost damage. All these climatic limitations are primarily related to the output from the sun – sunlight – though the occurrence of frost pockets depends on relief and weather conditions.

All crops need water to survive. Some crops require less water than others; some crops are better able to extract water from the soil than others. Crops take in most of their water requirement through their root systems. This means there must be in the soil a supply of water which can be tapped. The level of the soil water reservoir (how full it is) influences the ease with which plants can draw water off: when the level is low, crops have a hard job getting water out, and when a critical level is reached, they give up the ghost and wilt; when the level is high, there is too much water for most crops to handle and, perhaps surprisingly, they may wilt, not for want of water, but for an excess of it. The level of the soil reservoir is clearly crucial to crop growth. The amount of water stored in the soil depends partly on soil texture and partly on the extent to which losses in evaporation, plant uptake, and deep drainage (which takes water beyond the reach of crops) are made good by rain. Heavy rains and hail can damage mature cereal crops. Hail is a menace in the vine-growing areas of southern and central Europe, destroying large areas of nearly ripe vines. Heavy rain may also promote rill, gully, and sheet erosion of the soil cover, as well as leading to the flooding of fields and hence serious loss of crops in low-lying areas. Of importance to the agriculturalist are the amount and frequency of rainfall. In Britain, the heavy and frequent rain which falls in the west of the region makes conditions hazardous for harvesting crops. Grass fares better, but the rapid loss of nutrients caused by the high rainfall can lead to a deterioration in the quality of grass and periodic reseeding is necessary. The lighter, less frequent rain which falls over eastern England is, in most years, sufficient to support tillage crops and, except on light soils, the growth of good grass. The smaller number of rain-days improves the chances of a successful harvest, though thunderstorms can damage standing crops. In the west of Britain, the soil water reservoir, as well as being well fed by heavy and frequent rains, loses less water through evaporation and transpiration than is the case in the east. Soils in the north and west seldom have their stores of water depleted to a seriously low level. In the east, the soil water reservoir not only receives less rain, it also loses more water through evaporation. During summer months especially, the reservoir of water in the soil can drop to a dangerously low level, low enough to impair the growth of crops. This deficiency can, however, be made up by the addition of irrigation water. On average, green crops grown along the east coast need irrigating nine years in ten; and over the whole of East Anglia and southeast England they need irrigating seven years in ten. The amount of irrigation water which needs to be added to the soil water reservoir varies with different crops: less is needed for cereals and sugar beet than for grass, vegetables, and potatoes.

Wind can limit crop growth. Strong winds may damage mature cereal crops. Cold, local winds, such as the Mistral which blows down the Rhône valley in France, can devastate crops. Hot, local winds, such as the Sirocco which blows over southern Italy and Malta, can ruin crops by drying them out. Wind can also produce crop failures by eroding the soil.

SOIL MOISTURE DEFICITS IN ENGLAND AND WALES

The pores in the soil can store water. This soil water reservoir may be tapped by crops and other plants. Not all the water stored in the soil is available to crops. Some of the water, the hygroscopic portion, is so strongly attracted to the soil particles that plant roots cannot create a force large enough to extract it. Water which, when the soil is saturated or water-logged, fills large pore spaces, structural cracks, and channels is superfluous and even inimical to crops save for those, such as rice, which are water-loving (hydrophytes). Another portion of the soil water reservoir, the portion that can be used by crops, is held in small pores and cracks in the soil by capillary forces, the same forces which draw water into a dry sponge and hold it there. When all the capillary pores are full of water the soil is said to be at field capacity, which condition usually exists a few days after a storm when the water in large pores and cracks has drained away. At saturation, the soil moisture content is 100 per cent. The percentage by volume of the pores occupied by water at field capacity varies with soil texture: it is about 42 per cent for a clayey soil and a mere 14 per cent for a sandy soil (Table 2.1). The capillary water may be slowly depleted by evaporation and by plant uptake and by slow drainage through the soil. When the capillary water has been reduced to a critical value, called the permanent wilting point, crops wilt. As is evident in Table 2.1, permanent wilting point in a clayey soil is reached when about 25 per cent by volume of the pores are filled with water; in a sandy soil the figure is about 4 per cent. The capillary moisture stored between permanent wilting point and field capacity is potentially available to crops. The size of this useful reservoir of water is measured by the available water capacity (Table 2.1) which is expressed, like rainfall, in centimetres or millimetres. Notice that heavy soils can hold a lot of water, but much of it is unavailable to crops. Light soils, on the other hand, store less water, but most of it is readily available. So the soil water reservoir in a light soil is quickly exhausted by crops, whereas, in a heavy soil, the water is taken up more slowly but the supply lasts much longer.

Table 2.1   Field capacity, permanent wilting point, and available water capacity in some soils

Soil texture	Soil wetness at field capacity (per cent)	Soil wetness at permanent wilting point (per cent)	Available water capacity (mm per 100 of soil)
Sand	14	4	15
Clay	42	25	17
Loam	30	13	17
Clay loam	34	28	18
Silt loam	39	16	19
Very fine sand	20	4	23

Note: The available water capacity is not simply the difference between the water content at field capacity and the water content at permanent wilting point – the bulk density of the soil comes into the calculations.

Source: Simplified with permission from *Water, Soil and the Plant* by E. J. Winter (1974) published by Macmillan, London and Basingstoke, table 3.3.

The force by which water is held in the soil is highest for hygroscopic water and decreases gradually towards field capacity and beyond. The higher the force of retention, the more difficult it is for crops to withdraw soil water. As soil moisture falls short of field capacity, crops experience difficulty in taking water into their roots and a water stress develops which can inhibit growth. In Uganda, a sugar cane crop harvested after 720 days can be expected to yield about 200 tonnes per hectare in years when water stress develops on just a few days; but in relatively dry years the yield can be halved.

In agricultural studies, a measure known as the soil moisture deficit is often employed; it is defined as the amount by which the current soil moisture status falls short of field capacity and is usually expressed, like available water capacity, in centimetres or millimetres. A soil moisture deficit of 5 millimetres would mean that 5 millimetres of rain or irrigation water can be added to the soil to return it to field capacity. In the British climate, a soil moisture deficit, if it develops, will do so in the summer. Assuming the soil is at field capacity in early spring, it will stay in that state so long as losses caused by evaporation and transpiration are met by rainwater and, on some farms, irrigation water. When evaporation exceeds rainfall, the water in the soil will start to be depleted and a soil moisture deficit will accrue. Once below field capacity, the evaporation rate drops below its potential value; evaporation is then said to proceed at its *actual* rate rather than its potential rate. Potential evaporation, which takes place from lakes, ponds, rivers, and wet soil, is the maximum possible rate of water loss by evaporation under prevailing climatic conditions. Actual evaporation is the same as potential evaporation over lakes and other wet surfaces, but where water is in short supply the actual rate of evaporative water loss is less than the potential evaporative loss. A study of monthly rates of rainfall and evaporation can indicate the likely growth of a soil moisture deficit.

1   Table 2.2 lists monthly rainfall, potential evaporation, and soil moisture deficit amounts for three regions in England. Plot the information as a graph (Figure 2.1). The data for the upland zone of the northern Pennines have been plotted for you.

2   What problems do you think large soil water deficits around the Wash may create for Fenland farmers? How do you think the farmers might combat these problems?

3   To what extent are your suggestions borne out by the pattern of irrigation need in England and Wales as shown in Figure 2.2c?

Table 2.2

Month	Region								
	Low-lying districts around the Wash			Upland zone of southern Pennines			Upland zone of northern Pennines		
	Rainfall (mm)	Potential evaporation (mm)	Soil moisture deficit (mm)	Rainfall (mm)	Potential evaporation (mm)	Soil moisture deficit (mm)	Rainfall (mm)	Potential evaporation (mm)	Soil moisture deficit (mm)
January	48	1	0	110	−1	0	170	−1	0
February	38	11	0	79	4	0	119	3	0
March	38	32	0	69	24	0	103	22	0
April	37	57	0	71	46	0	103	42	0
May	44	85	0	74	71	0	99	66	0
June	47	93	86	74	81	0	100	73	0
July	56	94	104	89	79	22	117	72	10
August	60	80	109	113	63	21	152	56	0
September	49	50	99	119	36	0	171	30	0
October	50	22	0	116	15	0	169	15	0
November	58	5	0	117	0	0	175	0	0
December	50	1	0	120	3	0	185	−3	0

Source of data: Climate and Drainage, M.A.F.F. Technical Bulletin No. 34. H.M.S.O., London, 1976.

78

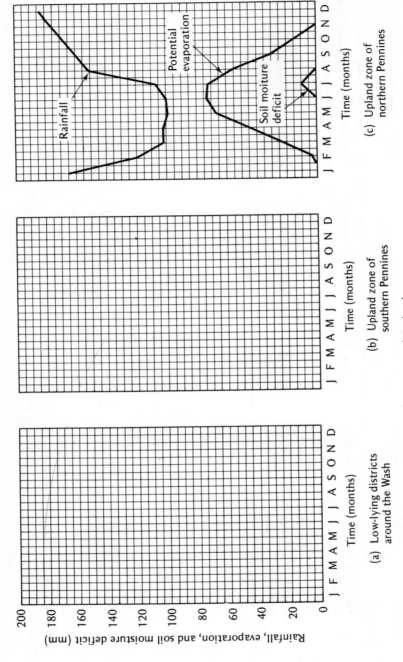

Rainfall, evaporation, and soil moisture deficit (mm)

(a) Low-lying districts around the Wash

Time (months)

(b) Upland zone of southern Pennines

Time (months)

(c) Upland zone of northern Pennines

Time (months)

Rainfall

Potential evaporation

Soil moiture deficit

Figure 2.1   The annual water balance in three parts of England

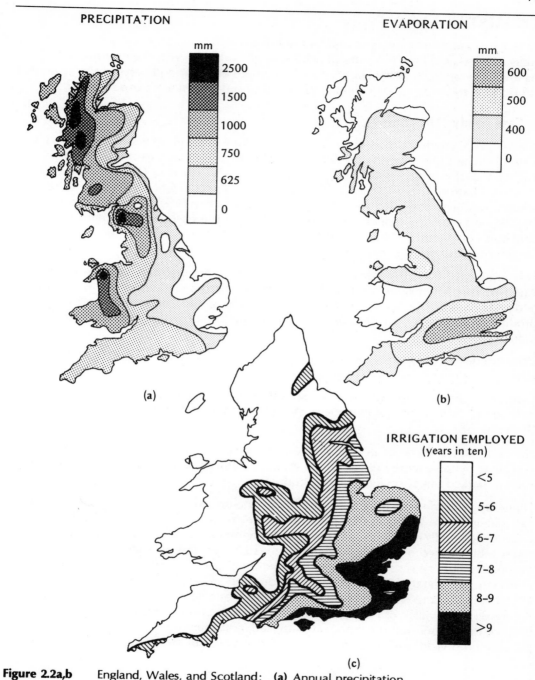

PRECIPITATION

EVAPORATION

mm
2500
1500
1000
750
625
0

mm
600
500
400
0

(a)

(b)

IRRIGATION EMPLOYED
(years in ten)

<5
5–6
6–7
7–8
8–9
>9

(c)

Figure 2.2a,b   England, Wales, and Scotland: **(a)** Annual precipitation
(b) Annual evaporation
Figure 2.2c   Theoretical irrigation need in England and Wales
Based on Penman's formula. Figures are for 1953, 1962, and 1966 returns.

IRRIGATION

By tapping water from rivers or aquifers to irrigate fields, it is possible to farm in areas of unreliable or inadequate rainfall. In large parts of Africa, for example, the expansion of agriculture is held back by a shortage of water and parts of the continent are unlikely to be developed so long as irrigation schemes remain expensive.

A Case Study: The Sudan

The Sudan is a semiarid country in which agriculture is severely restricted by a lack of water. North of Khartoum it is almost impossible to farm without some form of irrigation. Major irrigation schemes were started in the 1920s in Gezira, 100 kilometres south of Khartoum, and in the Gash delta, north of Kassala. Since then the Gezira scheme has been extended westwards towards the White Nile, bringing a further 300,000 hectares into agricultural use. Meanwhile, in the east, on the Atbara river, a dam was completed in 1965, the water from which has enabled another 200,000 hectares of land to be cultivated.

The dams have been expensive to build. The most recent one at El Roseires on the Blue Nile cost $89 million. But they have led to an increase in agricultural production and enabled a greater variety of farm produce to be grown. Since 1955, cotton production has more than doubled and the production of groundnuts has increased sixfold.

1   Why are dams so important in irrigation schemes in arid countries and how do they affect the annual pattern of streamflow in rivers?

2   Examine Figure 2.3 and suggest why major irrigation schemes have taken place in east-central Sudan and not elsewhere. What other information would help in answering this question?

Not all irrigation schemes in Africa have been as successful as those in the Sudan. The Niger delta scheme in Mali, which also dates from the interwar years, has not had such a marked effect on agricultural production.

There are two side-effects of irrigation. Firstly, the irrigation water may find its way to flat, alluvial plains and lead to a rise in the water table. The alluvial soils become waterlogged and crop yields are thus drastically reduced. Secondly, the rise in the water table caused by irrigation water may lead to salinization – the ground water evaporates but the salts dissolved in it do not, and they accumulate in the soil. Most crops cannot tolerate high salt concentrations and so crop yields are adversely affected. Both these side-effects of irrigation have affected 20,000 hectares of land in the lower Indus valley of Pakistan. The solution to the problem is to carefully balance the amount of irrigation water applied (and there must be enough to increase crop yield) with the amount of evaporation, but this is easier said than done.

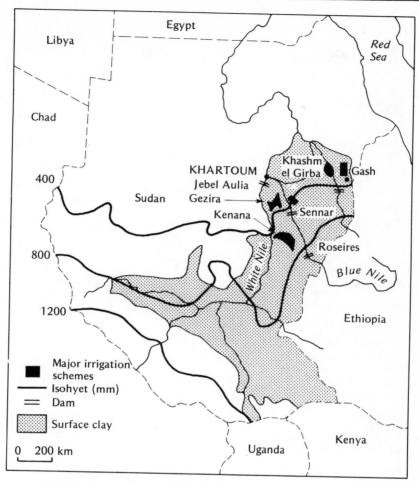

Figure 2.3   Some irrigation schemes in the Sudan

THE RELATIONSHIP BETWEEN LAND USE AND ALTITUDE AROUND LOUTH

Because climate changes with altitude, we might expect land use on hills and in valleys to differ. A glance at the map of land use west of Louth in Lincolnshire indicates that, by and large, pasture is found in valleys and arable land is found chiefly on hills (Figures 2.4a, b, and c). To test this notion, the area of study was, by superimposing a grid, divided into a number of square cells. Each cell was placed in one of three altitude zones – less than 200 feet, 200 to 300 feet, and more than 300 feet. Some 47 cells fell into the less-than-200-feet category, 60 cells into the 200-to-300-feet category, and 51 cells into the more-than-300-feet category. For each of the 158 cells, the dominant land use – arable or pasture – was noted (Table 2.3a). The results do suggest a relationship between land use and altitude zones, arable land being dominant on hills.

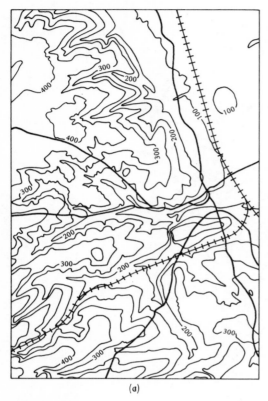

(a)

Figure 2.4a   Relief around Louth. Crown copyright reserved

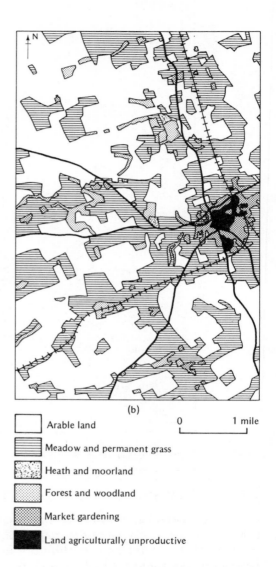

(b)

Arable land

Meadow and permanent grass

Heath and moorland

Forest and woodland

Market gardening

Land agriculturally unproductive

0   1 mile

Figure 2.4b   Land use around Louth. Data derived from the Second Land Utilization Survey

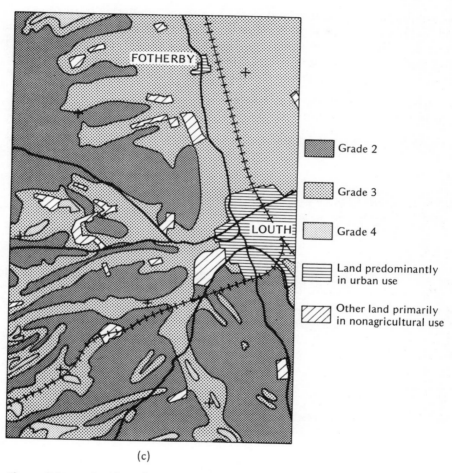

(c)

Figure 2.4c   Land grades around Louth
Source: Agricultural Land Classification of England and Wales, Sheet 105, MAFF.

To find out if the apparent relationship is borne out by statistical analysis, we can run a test which compares the observed frequencies of land use in each altitude zone with the frequencies we should expect to find were there *no* relationship between land use and altitude. The steps of this test are laid out in Table 2.3b.

Step one   The computation of expected cell frequencies

The expected frequency of arable land in the less-than-200-feet altitude zone is computed by multiplying the total number of cells in the zone (this is 47, the total of column one in Table 2.3a) by the total number of cells in all three altitude zones with arable land the dominant use (this is 103, the total of row one in Table 2.3a), and dividing the result by the total number of cells in the study area (this is 158). We have

$$\text{Expected number of cells in less-than-200-feet altitude zone dominated by arable land} = \frac{47 \times 103}{158} = 30.64$$

The remaining five expected frequencies are computed in like manner as shown in Table 2.3b.

Table 2.3a   Observed frequencies

Land use	1 Less than 200	2 200 to 300	3 More than 300	4 Row totals
	Altitude zones (heights in feet)			
1 Arable	18	39	46	103
2 Pasture	29	21	5	55
3 Column totals	47	60	51	158

Table 2.3b   Expected frequencies

Land use	1 Less than 200	2 200 to 300	3 More than 300	4 Row totals
	Altitude zones (heights in feet)			
1 Arable	$\frac{47 \times 103}{158} = 30.64$	$\frac{60 \times 103}{158} = 39.11$	$\frac{51 \times 103}{158} = 33.25$	103
2 Pasture	$\frac{47 \times 55}{158} = 16.36$	$\frac{60 \times 55}{158} = 20.89$	$\frac{5 \times 55}{158} = 17.75$	55
3 Column totals	47	60	51	158

Table 2.3c   Calculating the observed chi-squared value, χ^2_{obs}

Land use	1 Altitude zones (heights in feet) Less than 200	2 200 to 300	3 More than 300	4 Row totals
1 Arable	$\dfrac{(18-30.64)^2}{30.64}=5.21$	$\dfrac{(39-39.11)^2}{39.11}=0.00$	$\dfrac{(46-33.25)^2}{33.25}=4.89$	10.10
2 Pasture	$\dfrac{(29-16.36)^2}{16.36}=9.76$	$\dfrac{(21-20.89)^2}{20.89}=0.00$	$\dfrac{(5-17.75)^2}{17.75}=9.16$	18.92
3 Column totals	14.97	0.00	14.05	$\chi^2_{obs}=29.02{}^*$

* Notice that the observed chi-squared value is given by either the sum of the row totals *or* the sum of the column totals in this table.

Step two   The calculation of the chi-squared statistic

The next step is, for each of the six altitude–land-use combinations, to subtract the expected frequency from the observed frequency, square the result, then divide by the expected frequency. For arable land in the less-than-200-feet altitude zone we have

$$\frac{(\text{Observed frequency} - \text{Expected frequency})^2}{\text{Expected frequency}} = \frac{(18-30.64)^2}{30.64}$$

$$= 5.21$$

This procedure is carried out for all land-use–altitude combinations and the resulting six values are added up (Table 2.3c). The sum is the observed chi-squared value, written χ^2_{obs}, and in our example is 29.02.

Step three   Putting the result to the test

To decide whether to accept or reject the hypothesis that altitude and land use are *not* related (this is technically termed a null hypothesis since it assumes *no* relationship), the observed chi-squared value is compared with what is called a critical chi-squared value which can be read off from Figure 2.5. But to make the comparison we need to calculate what are called the degrees of freedom; these are defined for a table as the product of the number of rows less one and the number of columns less one. In our case, with two rows and three columns, the degrees of freedom are $(2-1)(3-1)=2$. We also need to define what level of statistical significance we are prepared to accept in affirming or rejecting the null hypothesis. We shall take the 99 per cent significance level, the line for which in Figure 2.5 intersects the two-degrees-of-freedom line at a critical chi-squared value, χ^2_{crit}, of 9.2. The observed chi-squared value, 29.02, is greater than the critical chi-squared value and so we can reject the null hypothesis that land use and altitude west of Louth are not related. We conclude, therefore, that, in 99 cases out of 100, land use *is* related to altitude by more than a chance association.

Is it really altitude, working through climate, which influences the pattern of land use? Considering the small range of altitude involved, we may reasonably raise doubts. How then can the relationship be interpreted? Is it just a spurious result thrown up by the chi-squared test?

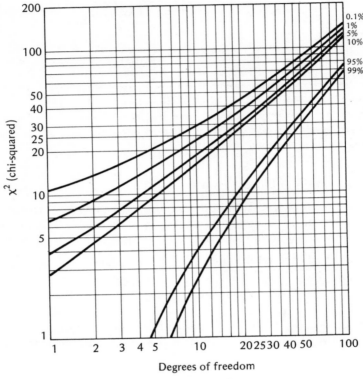

Figure 2.5   Graph for use in the chi-squared test.
The numbers on the right of the diagram refer to the lines and are probabilities. For example, the value 0.1 percent means that the association measured by the chi-squared value will occur by chance once in a thousand times; the significance level is thus 99.9 percent. The probability value of 1 percent corresponds to a significance level of 99 percent, the probability value of 5 percent with a 95 significance level; and so forth.
Reprinted with permission from *Statistical Methods and the Geographer* by S. Gregory (1978) published by Longman, London and New York.

1   Study the agricultural land-classification map of the study area (Figure 2.4c) and attempt to explain the relationship between land use and altitude around Louth. You may wish to attempt a chi-squared test, in which case you should superimpose a grid on the map. It should be evident that it is dangerous to make simple cause-and-effect statements about relationships between land use and physical factors.

Soil Limitations

The soil acts as a reservoir of nutrients for crops; only oxygen and carbon dioxide are exchanged directly with the atmosphere by way of leaves. Several nutrients are essential to plants and animals, and should any of them be in short supply production will suffer (Table 2.4). Nutrients needed in large amounts are known as macro-nutrients; those required in small doses are termed micro-nutrients.

Table 2.4   Nutrients essential to plants and animals

Macro-nutrients		Micro-nutrients	
Carbon	C	Iron	Fe
Hydrogen	H	Boron	B
Oxygen	O	Manganese	Mn
Phosphorus	P	Copper	Cu
Potassium	K	Zinc	Zn
Nitrogen	N	Molybdenum	Mo
Sulphur	S	Chlorine	Cl
Calcium	Ca	Sodium	Na
Magnesium	Mg	Cobalt	Co
		Iodine	I
		Chromium	Cr
		Gallium	Ga
		Vanadium	V

Nutrients circulate through natural and agricultural ecosystems alike. Agricultural man tampers with a natural nutrient circulation so that he might harvest a good portion of the nutrients (and energy) locked up in plant and animal tissues. One of the chief changes brought about by man is the harvesting of nutrients in leaves, roots, tubers, and fruits which would otherwise have been returned to the soil and recycled. The problem is to balance cropping, which removes nutrients stored in organic matter, with nutrient replenishment by fertilizer applications. If more nutrients are removed in the harvest than are added as fertilizer, then soil fertility will drop and nutrient levels may become a limiting factor to crop growth.

Apart from adding fertilizers to the soil to top up the nutrient reserves, a slow, natural replenishment by weathering of rock and as dissolved constituents in rainwater takes place. Where man is living at low population densities and practising subsistence agriculture, cropping losses are made good by natural replenishment. Swidden or slash-and-burn cultivation, practised by many peoples in tropical forests, depletes the soil's nutrient store. Given time, say fifteen to twenty years, the soil will recover naturally and may be reused; the soil fertility will not in the long term run down. If more frequent cultivations are made, soil fertility never quite reaches its previous level and gradually declines (Figure 2.6).

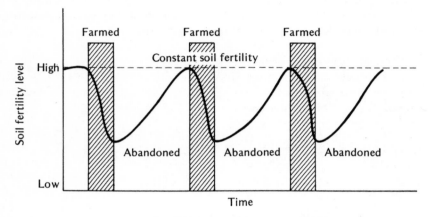

(a) Long recovery period – soil fertility level maintained

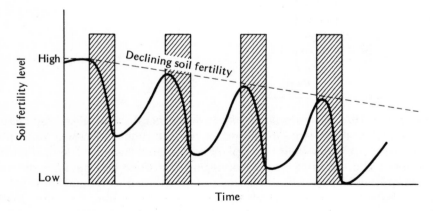

(b) Short recovery period – soil fertility level declines

Figure 2.6   The influence of slash-and-burn agriculture on soil fertility

In areas where man lives at higher densities, there is not time to let soil fertility recover naturally. Instead, nutrients are added to the soil in the form of fertilizers, either manure or synthetic ones. Fertilizers need to be used intelligently though.

Cations and anions of nutrients such as nitrogen, sulphur, phosphorus, potassium, and calcium, both those dissolved in the soil water and those attached to clay and humus particles, are available to plants but are also subject to washing away by rainwater, a process termed leaching. Nitrogen and sulphur anions are the most easily lost of nutrients. Only some 40 per cent of these nutrients added in fertilizers will be taken up by plants. Phosphorus, usually the first nutrient that must be added to the soil, tends to change to forms which plants cannot use, up to 75 per cent of applied fertilizer being rendered unusable in this way; leaching losses are usually small, however, and long-cultivated soils have good reserves of phosphorus.

LAND USE AND LAND CAPABILITY NEAR EASTNOR, HEREFORDSHIRE

With the exception of carbon dioxide and oxygen, the soil stores all the nutrients needed by crops. The amounts of nutrients stored in the soil, as well as other factors such as acidity which affect crop growth, vary with soil type. Different crops have different tolerances to soil properties. We might expect, therefore, that crops will tend to be grown on the most suitable soils for them; where they are grown on a range of soils, they will fare best, and so give the best yields, on soils best suited to their particular needs.

The smallest unit of soil mapped in the field is the soil series, every series being named after a place in the area it was first described. So we have the St. Albans series, the Southampton series, the Denbigh series. Soil profiles in each soil series will be found in a similar parent material (sandstone, alluvium, chalk, or whatever) and will have like properties. To test the hypothesis that soil type is related to land use we might compare the distribution of the soil in an area with the pattern of land use. A difficulty here is that, strictly speaking, all other factors relating to climate, economics, and the behaviour of the farmers should be held constant – not an easy task. A study which established a relationship between soil type and land use might well hide the complex interplay of factors which influence the agricultural system. And because of this, broad cause-and-effect associations between man and soil, once discussed at length in textbooks, can be exceedingly misleading. It is perhaps more justifiable to seek, in a small area in which economic factors at least are roughly uniform between farms, a relationship between an overall measure of land capability and land use. Land capability would at best reflect all key physical factors, such as soil texture, soil drainage class, soil acidity, length of growing season, soil moisture deficit, likelihood of frost, exposure, and many others. In practice, land-capability classes are based on a limited number of factors which are thought to have special significance to agricultural practices. The scheme devised by Bibby and Mackney of the Soil Survey of England and Wales is summarized in Table 2.5.

Table 2.5   Land-use capability classes

Class	Land limitations	Range of crops
1	Very minor or none	Full range
2	Minor	Some root crops and winter-harvested crops may be difficult to harvest
3	Moderate	Grass, cereals, and forage crops fare best
4	Moderately severe	Mainly grass; cereals and forage crops possible but rather a gamble
5	Severe	Pasture, forestry, and recreation
6	Very severe	Rough grazing, forestry, and recreation
7	Extremely severe	Very poor rough grazing for a few months; no forestry

Source: after the Soil Survey of England and Wales.

The kinds of limitation in each class are indicated by appropriate letters – w for wetness, s for soil limitations, g for slope and soil pattern limitations, e for liability to erosion, and c for climatic limitations. So on the map, 2w means class 2 land with limitations due to wetness which might be caused by slowly permeable materials like clay, high watertables, or flooding by rivers. Soil limitations include shallowness, which restricts root growth and nutrient uptake, and reduces the water-holding capacity of the soil; stoniness, which limits the water storage, reduces the nutrient status, and interferes with mechanical harvesting techniques; and adverse texture and structure, which may lead to reduced crop yields. Mechanized farming is difficult on steep slopes and complex patterns of soil, patches of bog in fields, for instance, may limit cultivation. Some areas are prone to wind and water erosion (this will be dealt with fully in the next chapter). Climatic limitations are assessed by the water balance, both average rainfall and average potential evaporation, and the long-term average of mean daily maximum temperature, to give three groups (Table 2.6).

Table 2.6   Climatic classes used in land-capability assessments

Climatic class	Residual rainfall (Average annual rainfall − Average annual evaporation) (mm)	Mean daily maximum temperature (°C)	Limitations on crops
1	Less than 100	Greater than 15	Few, if any
2	100 to 300	14 to 15	Moderately unfavorable to crop growth and choice of crops
3	Greater than 300	Less than 14	Severe limitations on choice of crops

Source: after the Soil Survey of England and Wales.

Figure 2.7a is a land-use capability map of part of the Eastnor Estate in eastern Herefordshire; Figure 2.7b is a land-use map of the same area. To test the hypothesis that land use on the estate is related to land capability, we shall adopt the following procedure.

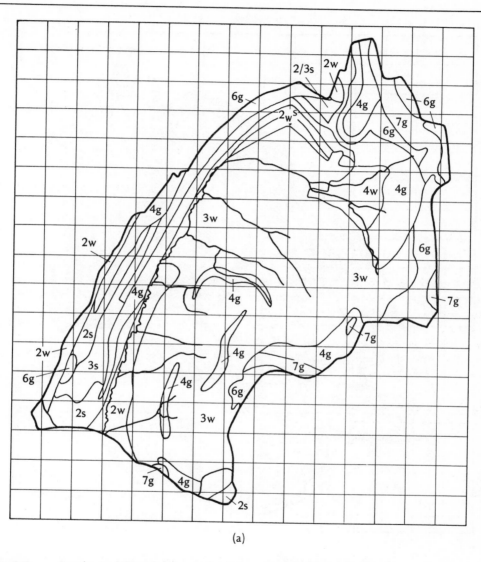

(a)

Figure 2.7a   Land capability classes on the Eastnor Estate, Herefordshire
Reprinted with permission from 'The use of soil maps in education, research and planning' by A. Warren and J. Cowie (1976). *Welsh Soils Discussion Group Report No. 17*, figure 3.

First of all, draw up a table in which column headings are land-use categories (use only three – oak and birch woodland, pasture, and arable and leys), and row headings are land-capability classes (use only four – class 2, class 3, class 4, and classes 6 and 7 combined); this is done in Table 2.7.

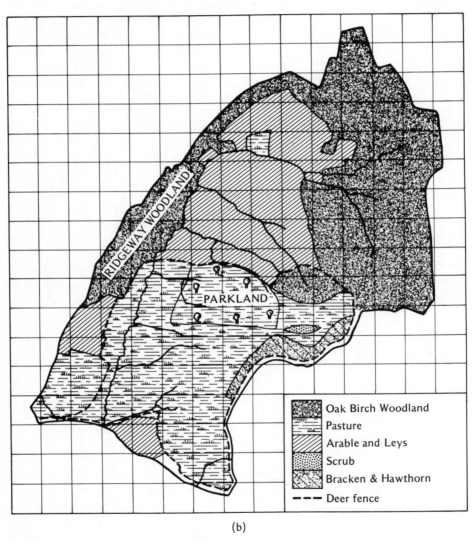

(b)

Figure 2.7b   Land use on the Eastnor Estate, Herefordshire
Reprinted with permission from 'The use of soil maps in education, research and planning' by A. Warren and J. Cowie (1976). *Welsh Soils Discussion Group Report No. 17*, figure 4.

Table 2.7

Land-capability class	Land use		
	Oak and birch woodland	**Pasture**	**Arable and leys**
2	✓✓✓	✓✓✓✓✓✓✓	✓✓✓✓
3	✓✓✓✓✓ ✓✓✓✓✓ ✓✓✓✓✓ ✓	✓✓✓✓✓ ✓✓✓✓✓ ✓✓✓✓✓ ✓✓✓✓✓ ✓✓✓✓✓	✓✓✓✓✓ ✓✓✓✓✓ ✓✓✓✓✓ ✓✓✓✓✓ ✓✓
4	✓✓✓✓✓ ✓✓✓✓✓ ✓✓✓✓✓	✓✓✓	
6 and 7	✓✓✓✓✓ ✓✓✓✓✓ ✓	✓✓	

Secondly, lay a grid over Figure 2.7a and over Figure 2.7b. For each grid intersection, note the land-use and land-capability class and record it by a tick in the appropriate box of Table 2.7. For example, a grid intersection which showed pasture and class 4 to be associated would be recorded by putting a tick in row 3 and column 2. Next, when the data for all the grid intersections within the study area have been recorded on the table, sum the ticks in each box and enter the results in Table 2.8.

Table 2.8

	Oak and birch woodland	Pasture	Arable	Row totals
Class 2	3	8	5	16
Class 3	16	25	22	63
Class 4	15	4	0	19
Classes 6 and 7	11	2	0	13
Column totals	45	39	27	111

This table contains the basic information we need and shows the number of grid intersections for which a given land use is associated with a given land-capability class. For instance, for the data shown, class 2 land is found to be associated with pasture at 8 grid intersections.

We are now in a position to see if any pair of land-use categories and land-capability classes is significantly related. Take the example of pasture and class 3 land. A new table, Table 2.9a, is drawn up from the data in Table 2.8.

Table 2.9a

		Pasture			Row totals	
		Present		**Absent**		
Class 3	Present	a	25	b	38	$a + b$	63
	Absent	c	14	d	34	$c + d$	48
Column totals		$a + c$	39	$b + d$	72	n	111

The data in this table contain four main items which are denoted by the letters a, b, c, and d: a is the number of grid intersections at which pasture and class 3 land occur together; b is the number of grid intersections at which class 3 land is present but pasture is absent; c is the number of grid intersections at which class 3 land is absent but pasture is present; and d is the number of grid intersections at which both class 3 land and pasture are absent. Also shown in Table 2.9a are row totals ($a + b$ and $c + d$), column totals ($a + c$ and $b + d$), and the grand total (n) which should be the same as the total number of grid intersections in the study area. Table 2.9a is termed a contingency table and from it a chi-squared value can be computed using the following formula

$$\chi^2 = \frac{(ad - bc)^2 \cdot n}{(a + b)(c + d)(a + c)(b + d)}$$

And, by comparing it with a critical chi-squared value for a given significance level at one degree of freedom (this is the degrees of freedom for a table with two rows and two columns), we can test the null hypothesis that pasture and class 3 land bear no relation to one another, that is, they are independently distributed within the study area. Putting the values from Table 2.9a into the chi-squared formula, we have

$$\chi^2 = \frac{\{(25 \times 34) - (38 \times 14)\}^2 \times 111}{63 \times 48 \times 39 \times 72}$$

$$= \frac{11,224,764}{8,491,392}$$

$$= 1.32$$

The critical chi-squared value at the 90 per cent significance level is 2.71 (Figure 2.5); since our observed chi-squared value falls below this critical level there is no reason to reject the null hypothesis and we may conclude that, in the part of the Eastnor Estate studied, pasture and class 3 land are distributed independently of one another – in short, they are unrelated to one another.

Test the relationship between:

1   i   class 2 land and arable land use;
    ii class 3 land and arable land use; and
    iii class 6 and 7 land and oak and birch woodland,

by completing the contingency tables (Tables 2.9b, c, and d), as they are called, and applying the chi-squared test to them.

2   What conclusion about the relationship between land-use capability and land use on the Eastnor Estate can you draw from your results?

Table 2.9b

		Arable		
		Present	**Absent**	**Row totals**
Class 2	Present			
	Absent			
Column totals				111

Table 2.9c

		Arable		
		Present	**Absent**	**Row totals**
Class 3	Present			
	Absent			
Column totals				111

Table 2.9d

		Oak and birch woodland		
		Present	**Absent**	**Row totals**
Classes 6 and 7	Present			
	Absent			
Column totals				111

SOIL TYPE AND AGRICULTURAL YIELD IN THE ROE VALLEY

In an attempt to see if there is a relation, as one would expect there to be, between soil type and agricultural yield, J. G. Cruickshank and W. J. Armstrong made a detailed survey of agricultural land in the Roe valley, County Londonderry, Northern Ireland. The Roe valley is a broad depression in which less resistant strata of Cretaceous, Triassic, and Carboniferous age, aligned in a north-south direction, have been eroded between Tertiary basalt in the east and Dalradian schist in the west (Figure 2.8). The rocks are overlain by glacial tills, the boundaries of which coincide

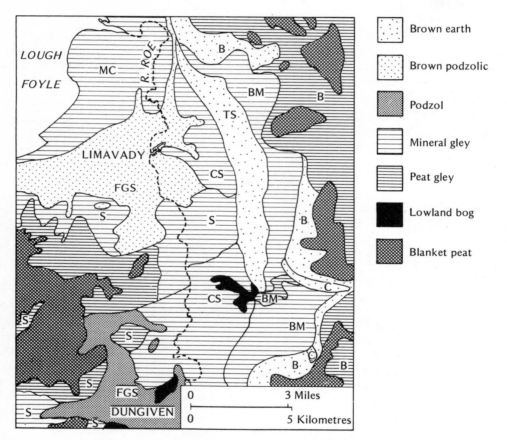

Figure 2.8   Simplified map of soil series in the Roe valley, County Londonderry.
The parent materials of the dominant soil profiles in each soil series are indicated with an initial letter: B = basalt till; Bm = mixed basalt and Keuper marl till; TS = Triassic sandstone till; CS = Carboniferous sandstone till; C = chalk; S = schist till; FGS = fluvioglacial sand and gravel. MC = marine clay (alluvium)
Reprinted with permission from 'Soil and agricultural land classification in County Londonderry' by J. G. Cruickshank and W. J. Armstrong (1971). *Transactions of the Institute of British Geographers*, 53, 79–94, figure 2.

almost exactly with their derivative rocks from which they inherit their textual physical properties. The mixing of till materials is evident only in the Triassic marl till which has an admixture of Cretaceous chalk and basalt. Extensive sand and gravel deposits of fluvio-glacial origin occur around Limavady and in other valley sites. In the Roe estuary is an area of reclaimed marine alluvium.

Seven soil series and the farms found on them were studied: marine alluvium gley; mixed basalt and Keuper marl till gley; schist till mineral gley; schist till peaty gley; Carboniferous sandstone till; sand and gravel podzols; basalt till gley. A stratified, random sample of ten farms within the size range 20 to 60 hectares and operated by men between the ages of 35 and 60 was taken from each of the seven soil series. All farms were roughly equidistant from the market nearest to them.

Data collected from the farms enabled gross margins per hectare for four main crop and livestock enterprises — oats, barley, potatoes, and livestock — to be calculated for each farm in the sample. Gross margins are given by the gross value of output less the production costs (for fertilizers, seed, and so on) per hectare of land.

The mean gross margins per hectare are given in Table 2.10 for each of the seven soil series. For each enterprise, Cruickshank and Armstrong applied a statistical test, known as analysis of variance, to see if the mean gross margins on each soil series differed significantly. The confidence levels at which the means for each soil series may be considered significantly different are listed in the bottom row of Table 2.10. The results show that, with the exception of oats, the yield of which is presumably indifferent to soil type, there are indeed significant differences in gross margins of the same enterprise on different soil series: soil type does influence agricultural yield in the Roe valley.

Table 2.10   Mean gross margins from sampled farms, grouped by soil series

Soil series	Oats (£/ha)	Barley (£/ha)	Potatoes (£/ha)	Livestock (£/ha)
Marine alluvium gley	None	61.2	None	32.6
Mixed till gley	32.5	None	184.5	42.7
Basalt till gley	None	44.0	176.3	58.7
Schist till mineral gley	36.0	52.3	226.5	26.5
Schist till peaty gley	31.2	None	260.5	31.7
Carboniferous sandstone till gley	37.0	54.2	230.0	49.0
Sand and gravel podzols	44.0	54.0	181.0	55.2
Per cent confidence level at which means for each enterprise may be considered significantly different between soil series	<95.0	97.5	99.9	99.0

Source: Reprinted with permission from 'Soil and agricultural land classification in County Londonderry' J. G. Cruickshank and W. J. Armstrong (1971). *Transactions of the Institute of British Geographers*, 53, 79–94, adapted from table III.

The Pattern of Farming in Britain

Economic and physical factors will not give a complete explanation of agricultural patterns. Other factors, including the pattern of land tenure, government action, and local labour availability may influence a farmer's preference for different enterprises. Indeed, as Alan Rogers put it, all farmers, to a greater or lesser extent, make decisions which are a compromise between economic and physical realities and their own whims and fancies. But whatever a farmer's preferences may be, some enterprises will better suit the ecological make-up of his land than others. And the ecological make-up of the land will eventually be reflected in the economic fortunes of the farmer, for the extra cost of working heavy land and of gaining low yields will diminish his profits. Because of this, local land-use patterns can reflect personal characteristics of farmers, but at a national level individual preferences by farmers are sorted and sifted by broad physical and economic constraints which come to dominate the picture. This is the case in Britain.

TILLAGE

Tillage refers to ploughed land which is generally under crops other than grass. The chief areas of tillage lie in the east (Figure 2.9a) where rainfall is less than about 750 millimetres a year and there are relatively few rainy days. Tillage is not important in western areas, particularly in the northwest, where high rainfall, low accumulated temperatures, and a lack of sunshine decrease the likelihood of a successful harvest. Some areas in the west do have relatively high proportions of tillage. In the southwest, temperatures and sunshine are more favourable to crop growth. Around coasts, as for instance in Pembrokeshire, frost is less likely and early crops can be raised. And some areas, such as the Eden valley and the Welsh borderlands, lie in the rain shadow of uplands.

Climatic conditions in the east are good for cereal growing (Figure 2.9b). Grass, on the other hand, is difficult to establish there and periods of drought in summer, which occur not infrequently, can severely retard grass growth. Arable crops are best grown on undulating land with light- and medium-textured, well-drained soils that can be worked at most times of the year. It so happens that such soils lie mainly in the eastern part of the country. Reinforcing these physical contrasts between west and east (Figures 2.10a and b) is the difference in farm size. Farmers in the principal tillage areas tend to have large farms and large fields, and have invested heavily in tillage equipment. Farmers in the west have smaller farms, less well suited to mechanized cropping techniques.

W. B. Morgan and R. J. Munton, writing in 1971, predicted that the contrasts between arable Britain and grassland Britain would be made stronger on joining the European Economic Community and adopting its Common Agricultural Policy. Cereals fetch high prices in Europe and so cereal production in Britain, particularly in eastern England, would increase at the expense of dairy farming.

The main cereals grown are wheat, barley, and oats (Figure 2.11). Wheat is mainly grown in areas with less than 1000 millimetres of rain a year, the ripening and harvesting of the crop being difficult elsewhere. Within areas of suitable climate, wheat, with its strong rooting system, is widely grown on heavy- to medium-textured soils (Figure 2.11a). Barley can be raised under many conditions but it prefers light, calcareous or at least nonacid soils — its distribution reflects these soil requirements (Figure 2.11b). Barley can be grown further north than wheat because it takes less time to mature. Oats are more tolerant of acid soils than wheat and barley are, fare well under cool and moist climatic conditions, ripen with a minimum of sunshine, but tend to suffer in dry seasons in eastern counties. So oats are mainly grown in the north and west (Figure 2.11c), though costs of production are on the high side and harvesting can be troublesome.

Root crops, which include potatoes, sugar beet, turnips, and swedes, are bulky, require a lot of labour, and are grown, for the most part, in relatively small hectareages. The physical requirements of root crops vary greatly. Potatoes, though they will grow under a wide variety of conditions, do best in deep, well-drained, stone-free soils of light to medium texture. The silts of the Fenlands and the soils of the Humber warplands fit the bill. The location of markets for potatoes influences the areas of

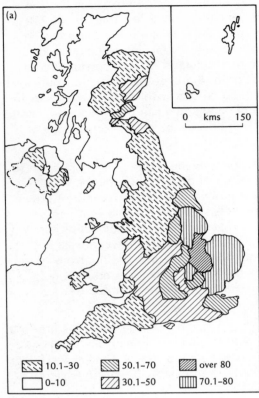

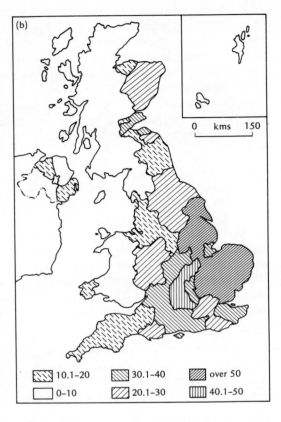

Figure 2.9a   Tillage, per 100 ha crops, grass, and rough grazing: United Kingdom, 1970 Reprinted by permission of Faber and Faber Ltd., from *Agricultural Resources: An Introduction to the Farming Industry of the United Kingdom* edited by A. Edwards and A. Rogers, 1974, figure 9.1.

Figure 2.9b   Cereals, per 100 ha crops, grass, and rough grazing: United Kingdom, 1970 Reprinted by permission of Faber and Faber Ltd., from *Agricultural Resources: An Introduction to the Farming Industry in the United Kingdom* edited by A. Edwards and A. Rogers, figure 9.2.

Figure 2.10a   The Tanat valley, Wales – part of the pastoral west. Fields are small, irregular in shape, and virtually everywhere under grass on which cattle graze. The open hill slopes are grazed by a few sheep. Reproduced with permission from Aerofilms

Figure 2.10b   The Fenlands, Lincolnshire – part of the arable east. Fields are larger and more regular in shape. Reproduced with permission from Aerofilms

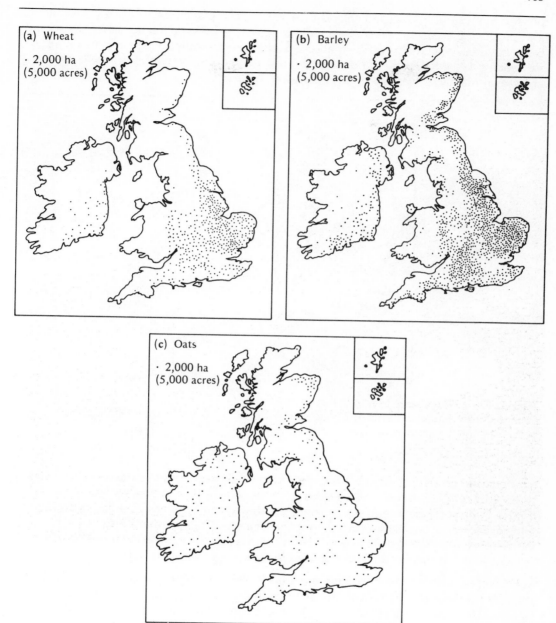

Figure 2.11 The distribution of **(a)** wheat, **(b)** barley, and **(c)** oats, 1970
Reprinted with permission from *The British Isles* (5th edition) by G. H. Dury (1973) published by Heinemann, London, figures 5.12, 5.13, and 5.14.

production (Figure 2.12): the Fenland area is centrally placed between several markets and producing areas in Yorkshire, Lincolnshire, north Cheshire, Staffordshire, Kent, and Essex have easy access to urban markets. Turnips and swedes, which are mainly used for feeding livestock, thrive under cool, moist conditions, preferably on light soils, and are grown mainly in western and northwestern districts.

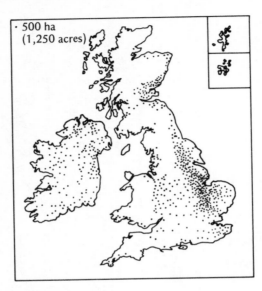

Figure 2.12   The distribution of potatoes, 1970
Reprinted with permission from *The British Isles* (5th edition) by G. H. Dury (1973), published by Heinemann, London, figure 5.16.

Horticultural crops include a wide variety of fruit, flowers, and vegetables. They tend to be grown in just a few areas (Figure 2.13). Vegetable production, for instance (Figure 2.14), is found in eastern England, notably in the Fenlands and Bedfordshire, and to a lesser extent in southwestern Lancashire and the Evesham district of Worcestershire. Orchards are restricted in the main to southern England, in particular to Kent and the Worcestershire and Herefordshire region (Figure 2.15).

1a Examine Figure 2.16a, which shows the areas in Britain where orchards occupy more than 20 hectares per 1000 hectares of crops and grass. By comparing this figure with Figures 2.16b and c, which show, respectively, the mean daily bright sunshine and the mean annual rainfall for Britain, describe the influence climate seems to have on the location of orchards.

 b What other factors do you think might influence the location of orchards? Explain how the factors you suggest would operate.

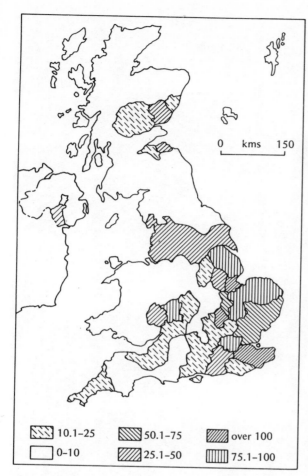

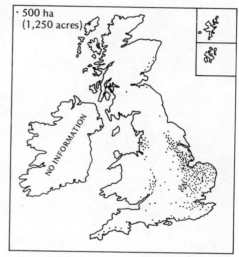

Figure 2.14 The distribution of commercial vegetables, 1970
Reprinted with permission from *The British Isles* (5th edition) by G. H. Dury (1973) published by Heinemann, London, figure 5.18.

Legend for Figure 2.13:

▨ 10.1–25   ▧ 50.1–75   ▨ over 100
☐ 0–10   ▨ 25.1–50   ▥ 75.1–100

Figure 2.13 Horticultural crops, per 1000 ha crops and grass: United Kingdom, 1970
Reprinted by permission of Faber and Faber Ltd. from *Agricultural Resources: An Introduction to the Farming Industry in the United Kingdom,* edited by A. Edwards and A. Rogers, figure 9.3.

Figure 2.15   Orchards and other fruit crops near Yalding, Kent. In the middle-distance are hop fields and the farm in the left foreground has five oast houses which were once used for drying hops. Reproduced with permission from Aerofilms

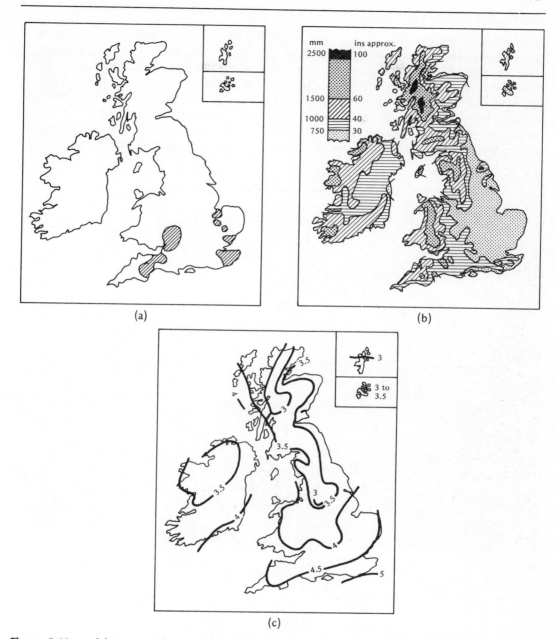

(a)

(b)

(c)

Figure 2.16 **(a)** Areas with more than 20 hectares of orchards per 1000 hectares of crops and grass (1958 census)
     (b) Mean annual precipitation
     (c) Mean daily duration of bright sunshine (hours)
(b) and **(c)** are reprinted with permission, from *The British Isles* (5th edition) by G. H. Dury (1973) published by Heinemann, London, figures 3.13 and 3.11.

GRASSLAND, CATTLE, AND SHEEP

Grass is the most important crop grown in Britain (Figure 2.17). It provides grazing for cattle, sheep, and the few remaining horses, as well as conserved feed as hay, dried grass, or silage. On climatic grounds, grass is of all crops the best suited to much of the country, preferring conditions which are not too wet (less than 1000 millimetres per year), not too dry (more than about 700 millimetres per year), have fairly frequent and well-distributed rainfall, and a long season of grass growth. Eastern and southeastern England, being liable to summer drought and containing large areas of light to medium soils, are not especially favourable to grass farming. However,

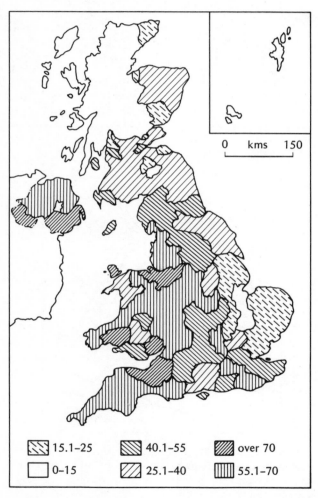

15.1–25   40.1–55   over 70
0–15   25.1–40   55.1–70

Figure 2.17   Grassland, per 100 ha crops, grass, and rough grazing: United Kingdom, 1970 Reprinted by permission of Faber and Faber Ltd., from *Agricultural Resources: An Introduction to the Farming Industry in the United Kingdom*, edited by A. Edwards and A. Rogers, figure 9.4.

heavy-textured soils, as in the Weald, and high watertables, as in Romney Marsh, can offset the disadvantage of low rainfall. A belt of land running from Northumberland to Dorset is climatically most suitable for the growth of good-quality grass, at least where soils permit. In the hilly western districts, where high rainfall tends to give rise to acid soils and water-logging, the best quality grasses cannot be grown but, especially in the southwest where the growing season is long, grass is widespread. The Welsh borderlands support good grass because rainfall is lower there than is usual in western areas.

The distribution of cattle, both beef and dairy, reflects the distribution of grassland (Figure 2.18a). Major concentrations are found in the west, notably the northwest and southwest of England, Northern Ireland, and southwestern Wales. Milk is the single most important product of farming in Britain. The chief concentrations of dairy cows are the climatically suitable Cheshire plain, Somerset and Dorset, southwestern Wales, and, to a lesser extent, Ayrshire and the Clyde valley (Figure 2.18b).

The present pattern of dairying can be understood only in an historical setting. Before rail transport, today's main dairying areas were national suppliers of butter and cheese. Indeed, most milk was made into butter and cheese or fed to livestock. Milk for liquid consumption was made in town dairies and in areas very near large urban markets, as Middlesex was. Reduction in the cost of transporting milk and technical improvements in methods of carrying it led to the gradual expansion of the area supplying urban markets with liquid milk. Areas like Buckinghamshire, which had formerly produced butter and cheese, turned over to liquid milk production; and dairying entered areas where it had never before been carried out. In 1933, the Milk Marketing Board was set up and fixed prices for milk which were the same for individual producers regardless of whether the milk was to be used for liquid consumption or to be made into dairy produce. This tended to reduce the contrast between areas remote from urban markets, where milk was made into butter, cheese, or other dairy products or else fed to livestock, and areas supplying liquid milk to urban markets. But even today, it is the remoter areas which supply most of the milk that is made into butter and cheese and other dairy products. The Milk Marketing Board also fixed at a very cheap rate the regional transport charges for the collection of milk and its carriage to depots or buyers; this has tended to diminish the importance of nearness to urban markets in explaining the areas of liquid milk production. Today, most of the milk from farms is sent to large collecting centres within the dairying regions and then taken by bulk transport to urban markets. The pricing policy has thus tended to make the physical suitability of an area for producing milk a more weighty consideration for the dairy farmer than nearness to an urban market. The chief dairying areas are now suppliers of milk for the urban population. This is the reverse of the situation that obtained in the mid-nineteenth century. Londoners, for instance, used to drink locally produced milk, but they now drink milk from the dairy farms of the west.

1   Study Table 2.11, which gives a region-by-region breakdown of milk production in England and Wales in the years 1938 and 1969.

a   Calculate the milk production of each region as a percentage of the national milk production (that is, the sum of milk production for all eleven regions) for 1938 and

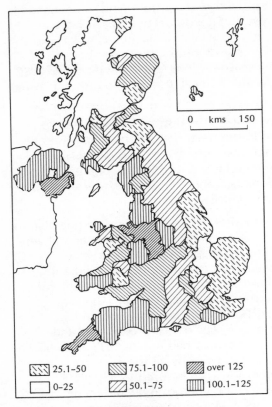

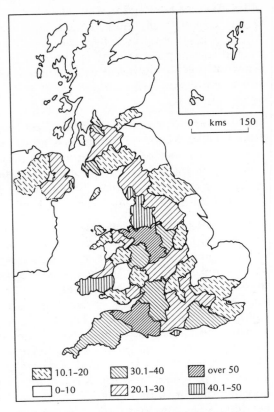

Figure 2.18a Cattle, per 100 ha crops, grass, and rough grazing: United Kingdom, 1970
Reprinted by permission of Faber and Faber Ltd. from *Agricultural Resources: An Introduction to the Farming Industry of the United Kingdom* edited by A. Edwards and A. Rogers, figure 9.7.

Figure 2.18b Dairy cows, per 100 ha crops, grass, and rough grazing: United Kingdom, 1970
Reprinted by permission of Faber and Faber Ltd. from *Agricultural Resources: An Introduction to the Farming Industry in the United Kingdom* edited by A. Edwards and A. Rogers, figure 9.8.

then for 1969. Calculate the change in the percentage share of national milk production in each region.

b Shade the regions in Figure 2.19a which have increased their percentage share in national milk production.

c To what extent does the intervention of the Milk Marketing Board explain the change in the pattern of milk production between 1938 and 1969?

2 Study Figure 2.20a, which shows areas of dairying in the USA classified by the main type of dairy product. With the help of Figure 2.20b, which gives an indication of the markets for milk products, explain the pattern of commercial milk production in the USA.

3 Study Figure 2.21 and, bearing in mind the physical environment of Britain, explain the distribution of sheep.

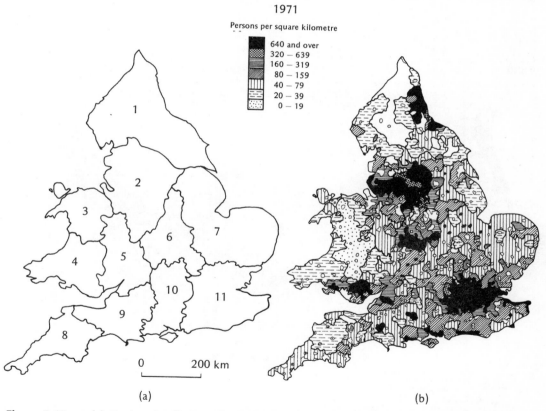

Figure 2.19 **(a)** Regional milk production in England and Wales: change in percentage share of national milk production between 1938 and 1969
(b) Population density in England and Wales, 1971
(b) is reprinted with permission from *The U.K. Space* (revised edition) edited by J. W. House (1977) published by Weidenfeld and Nicholson, London, figure 2.1.

Table 2.11 Milk production in England and Wales

	1938		1969		Change in percentage share between 1938 and 1969
Region	Volume (Megalitres)	Share of national output (per cent)	Volume (Megalitres)	Share of national output (per cent)	
1	340		990		
2	1355		2058		
3	145		490		
4	194		686		
5	436		980		
6	339		495		
7	240		392		
8	242		985		
9	726		1470		
10	335		588		
11	484		680		
Total	4836		9814		

COMMERCIAL MILK AREAS CLASSIFIED ACCORDING TO FORM IN WHICH
DAIRY PRODUCTS WERE SOLD FROM FARMS
(BASED ON COUNTY DATA)

Cream and farm butter areas
(whole milk sales less than one-third
of all dairy products sales)

Mixed areas
(whole milk sales less than two-thirds but more
than one-third of all dairy products sales)

Whole milk areas
(whole milk sales at least two-thirds
of all dairy products sales)

Noncommercial milk areas
(sales less than 6,500 pounds milk
equivalent per square mile)

(a)

Figure 2.20a   Areas farther from large markets sell more cream and butter than whole milk, since these concentrated products are relatively cheaper to transport.
From Earl O. Heady, *Economics of Agricultural Production and Resource Use,* © 1952. Reprinted by permission of Prentice-Hall, Inc., Englewood Cliffs, New Jersey.

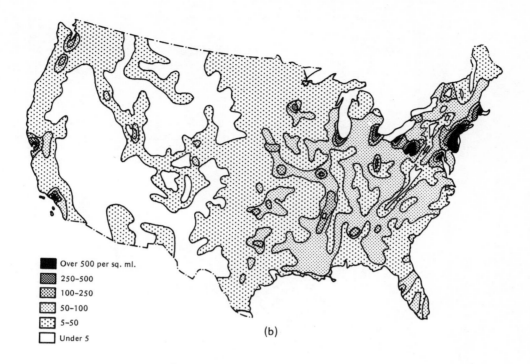

(b)

Over 500 per sq. ml.
250–500
100–250
50–100
5–50
Under 5

Figure 2.20b   Population density in the United States
Reprinted with permission from *A Social Geography of The United States* by J. Wreford Watson (1979) published by Longman, London and New York, figure 1.2.

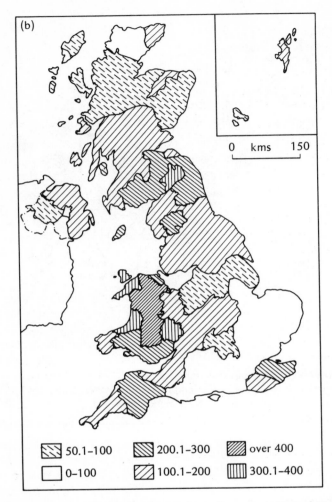

Figure 2.21 Sheep, per 100 ha crops, grass, and rough grazing: United Kingdom, 1970
Reprinted by permission of Faber and Faber Ltd. from *Agricultural Resources: An Introduction to the Farming Industry of the United Kingdom* edited by A. Edwards and A. Rogers, figure 9.10.

Biological Limitations

Energy from the sun is far and away the major source of energy which drives agricultural ecosystems. All crops – grass, trees, and so on – use sunlight, carbon dioxide from the air, and minerals and water from the soil to build plant tissues. The sunlight or solar energy is converted by the building process into chemical energy stored in crop tissue, in much the same way that muscles store chemical energy which is drawn upon when required. The complex process by which sunlight is converted into chemical energy is called photosynthesis and the essentials of it are captured in this equation

carbon dioxide + water + sunlight → sugar + oxygen + heat

Over a large area, say a field of wheat or an orchard, the making of organic matter by crops is called production rather than photosynthesis, but the process is the same. Because part of the photosynthetic process needs light, crops build reserves of chemical energy during daylight hours only. At night, the chemical energy store is used up by a process of slow oxidation called respiration for individual plants but consumption for large areas of plants. An example of respiration is

sugar + oxygen → water + carbon dioxide + energy

the energy so released being used to power the processes going on in the plant.

Crops, because they manufacture their own food from sunlight and raw materials, are termed autotrophs (from the Greek, meaning self-feeding). The amount of organic matter crops make during an interval of time is called gross primary production and is measured in units such as tonnes per hectare per year or grammes per square metre per day. Not all the organic matter produced by crops is available to humans or farm animals. Some 30 to 70 per cent of it is broken down by the crop to release energy which powers growth and other processes. This breakdown and energy release takes place in the process of respiration. Deducting the amount of organic matter used by the crop in respiration from the gross primary production, we arrive at the amount of organic matter produced by the crop which may be used by animals; this is called the net primary production. It usually runs at some 30 to 70 per cent less than gross primary production. In the English Breckland, wheat has a net production in the parts of the plant which stand above the ground of 5.2 tonnes per hectare per year. Figures for some other crops in the same area are 3.5 for barley, 9.0 for sugar beet, and 7.5 for grassland (hay), all values being tonnes per hectare per year.

The chemical energy of net primary production is available to cows, sheep, man, and other organisms which eat both living and dead plant tissue; and indirectly therefore is available to those animals such as foxes and eagles and the few plants which eat animal flesh. All these organisms which are dependent on other organisms for their food are called heterotrophs, or consumers. The storage of chemical energy in the tissues of heterotrophs is called secondary production. There are two broad groups of consumers. Firstly, the big ones, technically called macro-consumers or biophages, who eat living plant tissue; and, secondly, the tiny ones, the micro-consumers or saprophages, who slowly decompose the waste products (dung and urine) and the remains of other plants and animals.

Macro-consumers may be divided into the plant-eaters, also referred to as primary consumers or herbivores; and the flesh-eaters, also called secondary consumers or carnivores. In Europe, near the close of the last Ice Age, a large number of plant species of cold grassland, tundra, and boreal vegetation types were eaten by herbivores such as reindeer, mammoth, and arctic hare (Figure 2.22). In turn, the

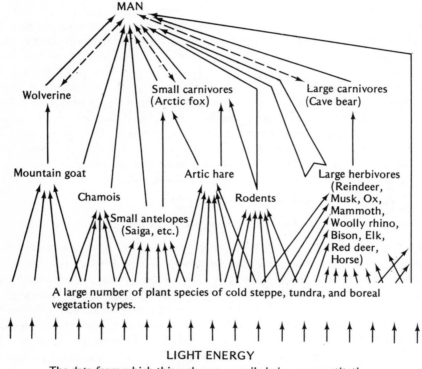

A large number of plant species of cold steppe, tundra, and boreal vegetation types.

LIGHT ENERGY

The data from which this web was compiled give no quantitative information concerning the abundance of the bone of the various animals, except to indicate the predominance of the large herbivore group in the bone assemblage.

– – – – – – Competitive relationship

———— Predatory relationship

Figure 2.22 Food-web diagram involving man (*Homo sapiens*) at the close of the last glaciation Reprinted with permission from *Biogeography: An Ecological and Evolutionary Approach* (2nd edition) by C. B. Cox *et al.* (1976) published by Blackwell Scientific Publications, Oxford, figure 70.

reindeer was eaten by the cave bear, a large carnivore, and the arctic hare was eaten by the arctic fox, a small carnivore. Man ate the arctic fox and in this role was a top carnivore or tertiary consumer. The simple feeding sequence, plant → herbivore → carnivore → top carnivore, is a grazing food chain; the example we have is tundra plants → arctic hare → arctic fox → man. Usually though, food and feeding relations in an ecosystem are far more complex than in a simple food chain because a wide variety of plants are commonly available, different herbivores prefer different plants, and carnivores are choosey about which herbivores they eat. Complications also arise because some animals, including man, eat both plants and animals; these are called omnivores. For all these reasons, energy flow through an ecosystem is better described as a food web.

Among the micro-consumers or decomposers, which live on dead material and waste products, there are also intricate food and feeding relations. A decomposer or detritus food chain (or web) may be recognized, the organisms in which are all micro-consumers. Plants, such as fungi, which live on dead material are saprophytes; animals which live on dead material, such as many bacteria, are saprovores. Dead organic matter and other waste products generally lie in or on the soil and, as they are decomposed by the micro-consumers, minerals are released which may be reused by plants; by this process the circulation of nutrients through an ecosystem is completed.

In a natural ecosystem then, energy is passed from one organism or group of organisms to another. Energy is gradually lost in respiration and excretion in its passage through the ecosystem at a rate of about 90 per cent loss at each step, and it must be constantly fed into the system, in contrast to nutrients which tend to be kept in circulation by the recycling of waste.

ECOLOGICAL EFFICIENCY

We have seen that crops produce organic matter of which about 30 to 70 per cent is used by the crops themselves in respiration. The remaining portion, the net primary production, is available to herbivores. In turn, herbivores produce organic matter but, being more active than crops, they use about 90 per cent of it in respiration; this leaves 10 per cent available to carnivores. Similarly, a mere 10 per cent of the organic matter produced by carnivores is available to top carnivores. The principle involved here is called ecological efficiency. It means that for every 1000 calories of plant material eaten by herbivores, just 100 calories are passed on to carnivores, and of these 100 calories, a trifling 10 calories reach top carnivores. This rapid loss of energy as food is passed along food chains explains why the number of links in a food chain seldom exceeds five. Clearly, from an agricultural point of view, it is best for man to grow and eat plants because this avoids the wasteful loss of energy in the conversion of plant tissue to herbivore flesh (see for instance, Table 2.12). However, if man wishes to remain a meat-eater as well, it would be sensible to farm herbivores as this avoids the wasteful loss of energy (90 per cent) in converting herbivore flesh into carnivore flesh. It is interesting that nearly all man's domesticated food animals are herbivores.

Table 2.12   Energy loss in two crops in the United Kingdom

Gross energy disposal of a **potato** crop in the UK:

Total organic dry matter formed per hectare	10,024 Megacalories	100%
Respiration loss	3,645	36
Unharvested vegetation	1,519	15
Postharvest loss	911	9
Household waste	911	9
Net human food	3,038	30

(which represents 0.22 per cent of the total solar energy received)

Gross energy disposal in an intensive grass crop grazed for **beef** production:

Total organic matter formed per hectare	11,340 Megacalories	100%
Respiration loss	3,645	32
Unharvested roots and stubble	1,215	11
Uneaten grazing	1,620	14
Dung and urine loss	1,620	14
Animal metabolism	2,025	18
Tissue conversion loss	405	4
Slaughter and household waste	203	2
Net human food	608	5

(which represents 0.02 per cent of the total solar energy received)

Source: based on data in *Farming Systems of the World* by A. N. Duckham and G. B. Masefield (1970) published by Chatto and Windus, London.

1   Examine the data in Table 2.13 and then describe and explain the relative efficiencies of the five farming food chains.

Not all herbivores have the same ecological efficiency, the 10 per cent being an average sort of value. Individual gross growth efficiencies, defined as the ratio of calories of growth to calories eaten, range from 4 to 40 per cent, the high values indicating efficient assimilation (taking up) of the food by the body, little being lost in faeces and respiration. Net growth efficiencies, defined as the ratio of calories of growth to calories of food assimilated, range from 5 to 60 per cent, the high values indicating that little of the assimilated food is lost as heat or respiration – most of it goes into growth. It is better to farm herbivores with high growth efficiencies. Or is it? Table 2.14 shows growth efficiencies for beef cattle reared on grassland and for a water flea (*Daphnia*).

2   Which of the two herbivores gives the more efficient conversion of solar energy into a food source for man? Why do you think beef cattle are at present raised in preference to the water flea?

When beef cattle are raised on grassland, they consume about one-seventh of all the grass available to them; the rest is eaten by other herbivores and decomposers which have no food value to man. So if it were possible to make sure the cattle get as much of the grass as possible, it would be possible to boost their productivity by

Table 2.13   Relative outputs of food chains in the United Kingdom

Farming food chain	Mcal of food per 100,000 Mcal of sunlight	Mcal of food per hectare	Edible protein (kg/ha)
Tillage crop→ man	225	2250	42
Tillage crop→ livestock→ man	23	260	68
Intensive grassland→ livestock→ man			
Meat	15	81	4
Milk	65	911	46
Grassland and crops→ livestock→ man			
Milk	40	324	17

Source: based on data in *Farming Systems of the World* by A. N. Duckham and G. B. Masefield (1970) published by Chatto and Windus, London.

Table 2.14

Herbivore	Growth efficiency (%)	
	Gross	Net
Beef cattle	4	11
Water flea	8	57

as much as seven times its present value. Modern intensive farming techniques achieve this for beef and other animals. One method is strip grazing: the cattle are tethered and not moved onto a new strip of grass until they have munched through all within the extent of their rope. Other techniques involve growing, harvesting, processing, and bringing grain and other foodstuffs to the animals, rather than letting them indulge in wasteful grazing, browsing, pecking, or whatever. The outcome of this is battery hens, calves raised in broiler houses, and cattle, both beef and milk, kept in what are called cowtels. There are a lot of moral issues involved here which you may like to discuss in class.

Further Reading

Agricultural Geography, W. B. Morgan and R. J. C. Munton, Methuen (1971), Chapters 3 and 4.
Pattern and Process in Human Geography, V. Tidswell, University Tutorial Press (1976), Chapter 4.
Farming in Britain Today, J. G. S. Donaldson et al., Penguin (1971).
We Plough the Fields, T. Beresford, Pelican (1975).
The Spatial Organization of Society, R. L. Morrill, Wadsworth (1970), Chapter 3.

CHAPTER THREE
THE ENVIRONMENTAL IMPACT
OF AGRICULTURE

The Impact on Soils

The toll of soil erosion, mainly induced by agricultural activities, on land resources is enormous. A detailed survey in the USA made in 1934 showed that, of a total of 167 million hectares of arable land, nearly 75 per cent was seriously damaged: 20 million hectares were ruined, 20 million were almost ruined, 40 million had lost more than half the topsoil, and another 40 million had lost a quarter of the topsoil. On the world scale, it is estimated that man has increased the world denudation rate from 20 million to 54,000 million tonnes per year. Almost all agricultural activities tend to increase the soil erosion rate because vegetation is cleared exposing the soil to attack by water, wind, and cattle. Despite the setting up of soil conservation services in many countries to overcome these problems, the erosion continues, usually because the immediate demand for food in a hungry world seems more important than the reduction of long-term productivity.

SOIL EROSION BY WATER

The key to increased water erosion in agricultural areas, especially arable ones, is the changes which occur in the water cycle when forest or grassland is laid bare or, at best, is given a patchy covering. Consider first what happens to rain falling on a wooded slope. A good portion of the rain will be intercepted by the leaves of trees and grasses, then drip from or run down leaves, stems, and branches until it reaches the ground. In a woodland, the upper soil layers are rich in organic matter, well aerated, and can soak up large quantities of water; they are said to have high infiltration capacities. Because of this, all but the heaviest of rains, unless the soil is saturated at the start of a storm, finds its way into the soil whence it either drains slowly into the ground waters or, more likely, seeps downslope to a stream as what is termed throughflow (Figure 3.1). Seldom does the upper soil become saturated to give rise to surface run-off (overland flow). On the other hand, in an open field there is no, or little, interception of falling rain by plants – the soil surface bears the full brunt of the falling rain drops. Every drop of rain that falls causes a spatter of surface soil; the overall effect of this is the gradual downslope movement of surface material. More importantly, as the surface soil contains little organic matter and is more compacted than in the woodland case, it has a lesser infiltration capacity; water, rather than seeping into the soil, tends to run over the land surface as overland flow

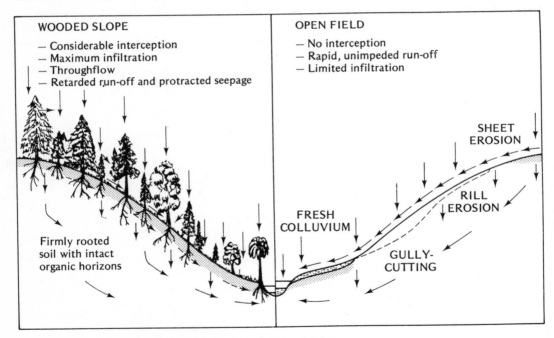

Figure 3.1   Run-off and infiltration on wooded and cultivated slopes
Reprinted with permission from *Geomorphology from the Earth* by K. W. Butzer (1976) published by Harper and Row, New York, figure 6.2.

and in doing so picks up soil particles dislodged by raindrops. The combined action of rain spatter and overland flow is known as sheet wash and it gradually removes topsoil on the slope. The eroded soil either accumulates at the slope bottom to form colluvium or passes into a stream. Overland flow may be concentrated along plough furrows or natural drainage lines to form rills, small channels, at most 1-metre wide and 50-centimetres deep, along which sediment is speedily transported (Figure 3.1). Gullies are much larger drainage features of arable land; they are channels, usually some 1- to 15-metres deep and wide, with steep or vertical sides and start at a permanent stream and eat back upslope through all soil horizons. Common the world over, gullies have many local names: *donga* in South Africa, *ravines* in French-speaking lands, and so on. Owing to their undesirable presence in agricultural land (unlike rills, tractors cannot negotiate them), they have been widely studied. They are certainly induced by overgrazing, but evidence of gullying episodes in the landscape before man's arrival suggests that this is not the sole cause; changes of climate which alter run-off rates and watertable levels may also be contributory factors. An interesting recent finding is that contour ploughing of slopes, a practice adopted to reduce run-off and sheet wash, increases throughflow which in turn leads to the incision of gullies; this illustrates the complexities of the land phase of the water cycle.

In detail, soil erosion by rainfall depends on two factors: the erosive power of the rain and the erodibility of the soil, each of which depends on other factors. The erosive power of the rain depends upon the quantity of rainfall, the annual distribution of rain, and the intensity of the rain (that is, the rainfall rate). On a world

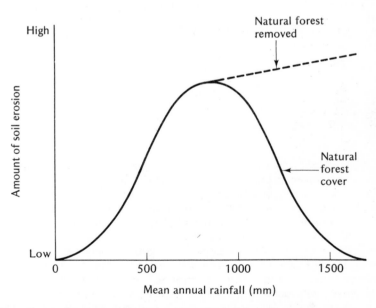

Figure 3.2   The relationship between mean annual rainfall and erosion
Adapted from *Soil Conservation* by N. Hudson (1971) published by Batsford, London, figure 1.3.

scale, the pattern of soil erosion caused by water is largely explained in terms of mean annual rainfall. The relationship is, however, a complicated one which is influenced by the type of vegetation cover. At low annual rainfall totals there is obviously little erosion. Large rainfall totals, say in excess of 1000 millimetres per year, are usually associated with dense forest vegetation which provides a protective cover, and so again erosion is not excessive. The greatest erosion tends to relate to medium rainfall totals, say, 500 to 1250 millimetres per year. However, if the natural cover afforded by forest in high rainfall areas is removed, soil erosion is severe (Figure 3.2).

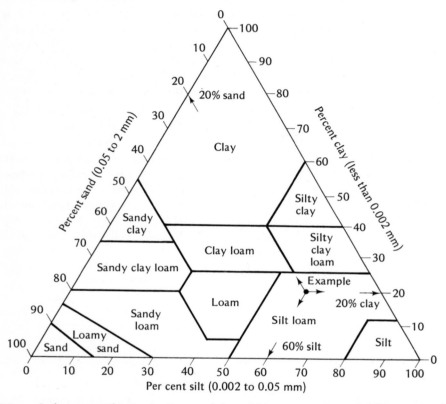

Figure 3.3   Soil texture. This is a measure of the relative proportions of different particle sizes in the soil. It is assessed by measuring the relative amounts of sand, silt, and clay in inorganic soil material that has been passed through a 2-millimetre sieve. There are several classifications of particle sizes: all concur that the upper size limit of sand particles is 2 millimetres and the upper size limit of clay particles is 0.002 millimetres (or 2 microns). To different combinations of the three size separates (sand, silt, and clay) are ascribed different names. The boundaries of the textural classes vary slightly, depending on how the distinction between sand and silt is made. One scheme is shown in this figure, which is known as a soil texture diagram or textural triangle. The use of the soil texture diagram is illustrated for the case of a soil sample which was found to contain 20 percent sand, 60 percent silt, and 20 percent clay. The point marked ● has the coordinate values corresponding to these percentages and lies in the 'silt loam' portion of the triangle. The soil would thus be referred to as a silt loam.

This broad pattern is further complicated by the annual distribution of rain – tropical areas tend to receive heavy rain during a wet season and are therefore more liable to erosion than temperate areas which generally have lighter rainfall spread more evenly throughout the year. The difference in tropical and temperate rainfall intensities reinforces this effect – in temperate areas rainfall intensities seldom exceed 75 millimetres per hour whereas in tropical areas intensities of 150 millimetres per hour are not uncommon. In semiarid areas, potentially the most susceptible to rainfall attack, rain falls in infrequent, short, intense showers which are capable of causing bad erosion. Temperate areas do not escape unscathed: steep slopes and certain vulnerable soils in these areas can suffer from soil erosion.

The second factor which will influence rainfall erosion, soil erodibility, is best appreciated on a local scale. Three things determine the erodibility of the soil. First of all, the physical make-up of the soil, its texture (Figure 3.3) and structure (Figure 3.4), is of prime importance in determining the ability of the soil aggregates to resist rain splash and overland flow. The relationships are, however, very complex and have been studied by soil scientists in depth. Bearing in mind the dangers of making general statements, it could be said that the more clay a soil contains, the more resistance it has to erosion because it has more cohesion and will tend to retain more moisture. Secondly, a vegetated surface being virtually immune to rain splash, and a bare soil being highly vulnerable, crop type may influence soil erodibility. Under permanent pasture, erosion rates are normal, but they increase under wheat and fallow, and are biggest with row crops such as sugar beet; a wise arable farmer will therefore manage his crops to ensure good ground cover at the times of year when erosion is most likely. Finally, soil erodibility is influenced by slope angle and slope length. The steeper a slope and the greater the length of a slope, the greater the potential for soil erosion. Where these two factors are a problem, the building of terraces in the landscape, as in paddy fields, or simple contour ploughing, can reduce the erosion.

SOIL EROSION IN EASTERN NORTH AMERICA: A CASE STUDY

In the forested parts of eastern North America, the clearance of forests and agricultural pursuits of pioneer settlers frequently led to severe soil erosion. The sediment stripped from the fields found its way into streams where it was deposited in large amounts. The recent geomorphological history of the Oak Ridges moraine area near Elizabethville, Ontario, in Canada has been described in detail by A. R. Hill. A survey carried out in the early 1840s showed the area predominantly covered by stands of oak and pine with one or two areas of improved land. In the decade from 1850 to 1860, settlement of the moraine took place and was accompanied by considerable forest clearance. Especially in the early period of settlement, farmers used a rotation of wheat and fallow, a combination which on the sandy, structureless soils of the area promoted erosion by water and by wind. By the end of the nineteenth century, soil erosion was bad enough for some areas to be abandoned. By the 1920s, the chief land use was permanent pasture and wasteland. Air photographs taken in 1927 revealed a severely eroded landscape: wind blow-outs ranging in size from a few

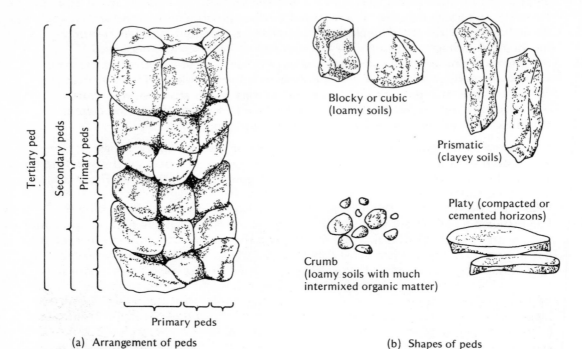

(a) Arrangement of peds

(b) Shapes of peds

Figure 3.4   Soil structure. The 'primary particles' of sand, silt, and clay which are found in the soil may, especially in sandy parent materials, be loose; in this case the soil structure is said to be single-grained. More commonly, the primary particles are held together by soil colloids ('gluelike' substances such as dehydrated oxides of iron and aluminium, calcium carbonate, and an organic material called humus). The bonding of primary particles may lead either to the formation of soil aggregates, or peds, separated by cracks; or, in the extreme case, to the formation of massive blocks which contain no cracks. Dig up a lump of soil and toss it to the ground and it will break into its component peds. Peds come in various shapes and sizes, some of which are illustrated in this figure.

square metres to ten hectares could be seen; steeper slopes showed signs of extensive sheet and rill erosion; and in the valleys, gulleys were actively forming. A detailed erosion survey conducted in 1942 reported widespread removal of 50 and 75 per cent of the topsoil, and in some areas the entire soil cover had been removed. Since the early 1950s, a reafforestation programme has greatly reduced soil erosion in the area.

A similar problem of soil erosion arose in the upper parts of the Tennessee valley. The upper catchment of the Tennessee river lies in the southern Appalachians. Pioneer farmers cleared the land by cutting and burning. They, and their descendants, planted cash crops like maize, tobacco, and cotton which made heavy demands on the reservoir of nutrients in the soil. With few livestock to produce manure and with no commercial fertilizers available, soil fertility dropped. Perhaps for ease of working, farmers ploughed up and down the slope of the land and planted cash crops in rows along the line of the furrows. During winter the soil lay bare. As a result, soil erosion set in. Topsoil was lost from farmland by sheet wash and by rilling and gullying (Figure 3.5). Remoter areas were put under the plough. But slopes in the remoter areas were steep and soil erosion was severe. Even on flat areas, the reservoir of soil nutrients dwindled fast and crop yields fell. Conditions on the farms became lamentable. Disease was rife: tuberculosis spread; so did malaria, transmitted by mosquitoes from the increasingly choked, increasingly swampy channels of the stream network; and so did typhoid, carried mainly in flood water, flooding having become far more serious a problem than in the days of the first settlers.

Figure 3.5   Soil erosion in the Tennessee Valley, USA. Reproduced with permission from Aerofilms

In 1933 the Tennessee Valley Authority Act was passed by the federal government. Its declared aims were to promote flood control on the Tennessee river system, to improve navigation, to establish hydro-electric power stations, and to achieve proper use of marginal farmland. Soil erosion is being combatted by the purchasing of some areas and putting them back under forest, by campaigns to persuade farmers to build check-dams in gullies, by the adoption of strip cropping, and by the sowing of grass and clover on some slopes to provide a year-round cover. All these measures help to reduce the charge of sediment into stream channels; they also tend to even out run-off, and so reduce to a minimum the risk of flash floods on the smaller streams (see *Settlements*, p. 192).

1 Look at the information given in Table 3.1, which shows typical variations of erosion under different types of vegetation cover, and then try to explain the changes in erosion rates in the Piedmont region of the United States as depicted in Figure 3.6.

Table 3.1

Vegetation	Soil erosion (mm/yr)
Forest	0.008
Pasture	0.030
Scrub forest	0.100
Barren, abandoned land	24.4
Crops (contour ploughing)	10.6
Crops (downslope ploughing)	29.8

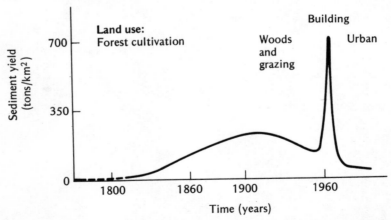

Figure 3.6   Changes in sediment yield of rivers with time in the Piedmont region of the United States
Adapted from 'A cycle of sedimentation and erosion in urban river channels' by M. G. Wolman (1967). *Geografiska Annaler*, 49A, 385–395.

WIND EROSION

Some 100 million tonnes of dust are injected into the air each year. Over 90 per cent of this total is derived from wind erosion of the land, the remaining portion coming largely from industrial emissions. On the world scale, desert areas are naturally subject to wind erosion; in a natural state, humid areas are virtually free from wind erosion. Once put under the plough, however, soils of humid areas are, under certain common conditions, vulnerable to attack by the wind. A classic example is the Dust Bowl of the central plains of North America which developed during the early 1930s and covered parts of five states – Colorado, Kansas, New Mexico, Oklahoma, and Texas. The natural vegetation of the plains was open prairie grassland adapted to low rainfall and occasional severe droughts. Settlers from the east, where rainfall is more plentiful, ploughed up the prairie and planted wheat. Wet years saw good harvests; dry years brought crop failures and dust storms, the worst of which were the disastrous ones of 1934 and 1935 which led to the abandonment of farms and an exodus of families. The land eventually had to be put back under grass. The effects of the dust bowl on cattle and man were bad indeed: livestock died from eating excessive amounts of sand; human sickness increased because of the dust-laden air; machinery was ruined, cars damaged, and some roads became impassable. The starkness of the conditions is evoked in a report of the time:

The conditions around innumerable farmsteads are pathetic. A common farm scene is one with high drifts filling yards, banked high against buildings, and partly or wholly covering farm machinery, wood piles, tanks, troughs, shrubs and young trees. In the fields nearby may be seen the stretches of hard, bare, unproductive subsoil and sand drifts piled along fence rows, across farm fields, and around Russian thistle and other plants. The effects of the black blizzards are generally similar to those of snow blizzards. The scenes are dismal to the passerby; to the resident they are demoralizing.

(Arthur H. Joel, *Soil Conservation Reconnaissance Survey of the Southern Great Plains Wind-Erosion Area*, 1937. *Note*: a black blizzard was a massive dust storm which blotted out the sun and turned day into night.)

Why should the American plains suffer such drastic wind erosion when cultivated, but remain erosion-free in the natural state? The answer is, on the face of it, simple: wind can erode a dry, unprotected soil only. A complete plant cover prevents the wind from touching the soil surface and eliminates the possibility of wind erosion; in an area of patchy plant cover erosion by the wind may take place.

The key to the prevention of wind erosion on agricultural land is the presence of a plant cover during danger periods (usually those times of year when the soil is dry and winds are strong). For instance, the practice today in those parts of the American plains where there is enough rain for the continuous cropping of a rotation of spring wheat and autumn rye is to seed each crop through the stubble of the previous one. A recent innovation is the covering of the soil, not with plants, but with protective films made from water-based emulsions of resin, asphalt, starch, latex, or oil, though the latter only on sand dunes, beach sand, and spoil heaps. Other, more subtle factors influence wind erosion and can be manipulated to control it. First of all, wind cannot pick up particles from a moist surface. If irrigation facilities are not available, the best way to keep the surface soil moist is to reduce evaporation; this may be done by adding organic matter, such as manure, to the soil, a process called mulching, or, in areas with unpredictable and generally low annual rainfall, by dry farming. The practice of dry farming, which is usually carried out with drought-resistant strains of plant, involves tilling the soil by contour ploughing to increase infiltration capacity and surface storage capacity, and thus reduce overland flow and evaporation. The same ground is then cropped in alternate years so that, weather conditions permitting, the soil water reservoir will be topped up during the fallow year. Secondly, the farther the wind blows unimpeded by trees, hedges, walls, and the like, the greater the potential the wind has for eroding the soil. Natural and artificial windbreaks aligned across the path of prevailing winds are thus helpful in alleviating wind erosion. Finally, rough surfaces, because they break up the wind flow and thus dissipate much of the kinetic energy of the wind, are less vulnerable to wind erosion than smooth surfaces. Surface roughness can be increased during periods in which wind erosion is likely by tilling (this has the added advantage of improving soil structure).

In Britain, the loss of soil by wind erosion, or blowing as it is colloquially known, is a worse problem than erosion by water. Areas most susceptible are the light, sandy soils of East Anglia, Lincolnshire, and East Yorkshire, and the light peats of the Fens. Records show blowing occurred as long ago as the thirteenth century, but the problem seems to have become severe in the last decade or so. Blows can remove up to two centimetres of topsoil containing seeds, damage growing crops by sand-blasting them, and lead to the blocking of ditches and roads. The increase in severity of blowing has been attributed to recent agricultural changes in the area: the change from using farmyard manure to inorganic fertilizers on fields, the use of heavy machinery to cultivate and harvest some crops, and the removal or grubbing of hedgerows to form larger fields better suited to modern, mechanized farming.

THE LOSS OF SOIL STRUCTURE

Good soil structure (Figure 3.4) is essential to efficient farming. Some agricultural soils in England and Wales have recently suffered a loss of structure and this has led to a drop in crop yields. The problem is not nationwide, the soils especially vulnerable being those developed on the Lias clays of the English Midlands, the clay soils of the London basin, the noncalcareous boulder clay soils north of the Wash, and some of the soils in the 'skirt' lands of the Fens where peat has been washed away (Figure 3.7). Three factors have contributed to the loss of soil structure.

Firstly, the cropping of the same cereal, particularly barley, year after year, has tended to reduce the store of organic matter in the soil. Organic matter is important in maintaining soil structure; without it, soil aggregates break up, as has happened in many areas where barley is grown. The loss of soil structure in this way can be remedied by draining, liming, and putting grass, which helps in the formation of a well-developed structure, in a rotation with cereals. Secondly, in areas which are intensively grazed, the grass tends to peel off, root depth becomes shallower, and the soil surface becomes poached during summer. And finally, ploughing can lead to a bad structure in the soil, this effect being a widespread problem on arable land. In a field ploughed year after year, a massive or platy layer develops at between 20 and 30 centimetres from the surface; this layer is called a plough pan and inhibits rooting and encourages water-logging of the surface soil (Figure 3.8). Left alone, soils with a plough pan recover after a few years, but drainage, busting, and subsoiling speed up recovery.

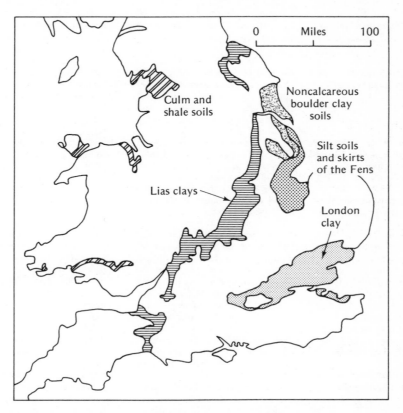

Figure 3.7 Soils subject to loss of structure in England and Wales

British soils under strain

By Leonard Amey   Reprinted with permission of *The Times*.

British soils over quite large areas of the country, are under strain. The past three years have demonstrated this only too clearly to the men who farm them.

The damage that has been done is seldom spectacular. One will rarely find anything impressive enough to make a photograph unless a hole is dug and a profile exposed. Talk of dustbowls and the kind of gulley erosion which took place in the United States during the 1930s is quite misplaced.

The trouble mainly is on soils that would hardly blow away in a hurricane and it is below the surface. It shows itself in difficult cultivations, a change in the weed flora and, most important of all to the farmer, static or falling crop yield.

The prophets of doom have been vocal enough for a long time now. The average working farmer has been apt to dismiss them as cranks—which, perhaps, some of them are.

For him economics have dictated a switch from mixed farming, using a good deal of labour, to straight cereal cropping, using little. His new machines have enabled him to work his land under conditions which his father would have found quite impossible.

Beneath the wheels of his heavy tractor changes were taking place of which he was usually quite unaware. They were slow and might have gone unnoticed for years yet, if the weather had not turned unkind.

In 1968 he was probably expecting a record cereal crop. It was smashed to pieces by the rain and wind. The autumn continued wet and the 1969 spring was atrocious. In places no sowing or planting was possible until the end of May. The autumn of last year was kind enough but this year's spring ran late and then a long drought set in.

Some farmers, on what were now recognized as problem soils, had three crop failures in a row and some of them have already gone out of business.

Last year a working party of soil scientists and farmers was set up and looked at the problem on the ground during the winter and spring. Their voluminous report is due for publication soon which the Minister of Agriculture, Mr. James Prior, promises should make good Christmas reading.

Some findings may well be anticipated during a conference at Bury St. Edmunds on Wednesday, arranged by the Soil Association, with Professor Sir Joseph Hutchinson in the chair. Soil structure, the importance of micro-organisms, the effect of heavy fertilizer dressings and practical ways of using organic material will be examined.

On the same day Fisons and farmers will discuss Cambridge problems of heavy land, including the economics of the break crops which might help the cereal men out of some of their soil troubles. It is, in fact, on the heavier soils that breakdown of structure has been most marked and some of it is due to ill-judged crop rotations.

The soils most vulnerable to modern misuse are the Lias clays of the English Midlands, the London clays, the non-calcareous boulder clays north of the Wash and some of the 'skirt' lands in the Fens where the peat has washed away. Also giving trouble are some silts and some of the Culm and shale soils of north-west England and Wales.

Over the centuries some of these have far more often been under grass than ploughed. The history of most acid clays has a common pattern. When corn prices are high (as during a war) they are ploughed and fertility built up under grass and stock is cashed. After a few years, falling crop yields combine with falling prices to send them back into grass. More recently fertilizers and chemical weed control may have delayed the drop in yield but have not prevented it.

Their drainage usually leaves something to be desired. It may be good enough for summer grazing but not for timely cultivation in the spring. If heavier and more powerful machinery is brought in to bully the land into some kind of tilth what structure remains is squeezed out.

The effect is worst on soils with little organic matter. A mainly cereal rotation, especially if barley plays a large part in it, will almost always lead to some drop in the organic content.

This may not seem to matter, because it will not show up in yields, until it reaches a certain critical point. Then a collapse of soil structure will make it impossible for the crop roots to function properly.

Even very light sands, when they are saturated in an abnormally wet spring, can collapse into something like wet concrete and dry out nearly as hard. But this is a relatively temporary effect, though if a crop was sown before the trouble started it will need to be redrilled.

The way back for the problem soils will usually involve attention to drainage, possibly liming, and cropping which puts back structure. What does this best is grass. On the silts which have given trouble even one year in grass will practically restore the position; on sticky, acid clays, probably, grass should largely replace arable cropping.

This is really an economic question, turning on the relation between cereal prices and those of milk, beef and lamb, it is easier to learn tractor driving than good grass management.

To get a comparable living from the second more animals have to be fed to the acre. Intensive grazing has led to other problems on lowland farms in the north-west and Wales, with Culm and shale soils.

At traditional rates of stocking they behave reasonably well. When this is stepped up the grass starts to peel off as rooting becomes shallower; pasture can be quite badly poached by the middle of July, since this is a country of high summer rainfall.

Some farmers seem, however, to have found a simple answer. They take a slitting implement on a long-spiked roller over their fields in June; the grass roots are able to function normally again and to use the water instead of drowning in it.

Texture

Structure

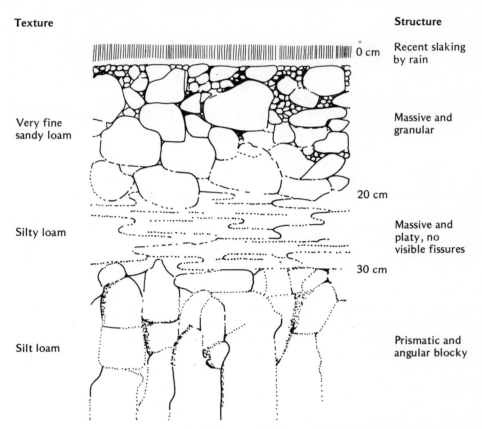

0 cm   Recent slaking by rain

Very fine sandy loam   Massive and granular

20 cm

Silty loam   Massive and platy, no visible fissures

30 cm

Silt loam   Prismatic and angular blocky

Figure 3.8   A typical, badly structured, silty loam profile, with faulty structures developed under arable cultivation. The massive and platy layer from 20 cm to 30 cm begins with a plough pan and is thickened by slaking and pressure. The topsoil is partly massive due to slaking. The lower horizon is well fissured.

The Impact of Pests and Their Control

Plant and animal species which interfere with man's activities are labelled as pests. Man's response to pests is to try to control the damage they cause. A control operation is successful if a pest does not cause excessive damage; it is a failure if a pest does cause excessive damage. Just how much damage is tolerable depends on the enterprise and the pest: an insect which destroys 5 per cent of a pear crop may be insignificant in ecological terms but it may be disastrous for the farmer's margin of profit; on the other hand, a forest insect may strip vast areas of trees of their leaves but the lumber industry will not go bankrupt.

Pests on agricultural land are commonly controlled by toxic chemicals called pesticides, which have an impact on the environment. In fact, pesticides offer a short-term solution to pest control; in the long run they are not efficacious. There are three reasons for this. In the first place, toxic chemicals can have adverse effects on animals and plants other than the pests. In her book *Silent Spring*, Rachel Carson drew attention to the alarming extent to which long-lasting pesticides were building up in soil, lakes, rivers, wild life, and man. One feature of pesticides, such as DDT, which are held firmly in animal tissue is particularly unpleasant: though applied in concentrations which are harmless to all but the pest, concentration levels build up as one organism eats another and the pesticide is fed along a food chain (Figure 3.9). Kenneth Mellanby, in his book *Pesticides and Pollution*, sites the example of an American lake into which DDD, a close relative of DDT, has found its way. The concentration of DDD in the lake water is 0.015 parts per million, but at successive stages of the lake food chain the values are 5 parts per million in plankton, 10 parts per million in small fish, up to 100 parts per million in large fish, and the fatal level of 1600 parts per million in birds which prey on the fish. Not all pesticides pollute the environment. Modern herbicides like *Dalapon* and *Paraquat* are short-lived, being rapidly neutralized by soil and water. Fungicides are longer-lived but, because they tend to be used sparingly, do not present the pollution problems they might. Mixtures of copper sulphate and other sulphur compounds are used to control blight on fruit trees and potatoes. Local concentrations of these compounds do not seem to do much harm and they are soon diluted to harmless levels by water. More serious is the use of organo-mercury compounds in seed beds. Mercury, like DDT, is persistent and is stored in body tissues. The high levels of mercury in tuna fish in some parts of the world have been traced back to mercury applied in seed dressings. As far as insecticides go, the group based on organo-phosphates, though many of them are exceedingly poisonous, do have the merit of quickly decomposing. The group based on organo-chlorines, of which DDT is a member, have proved efficient in controlling

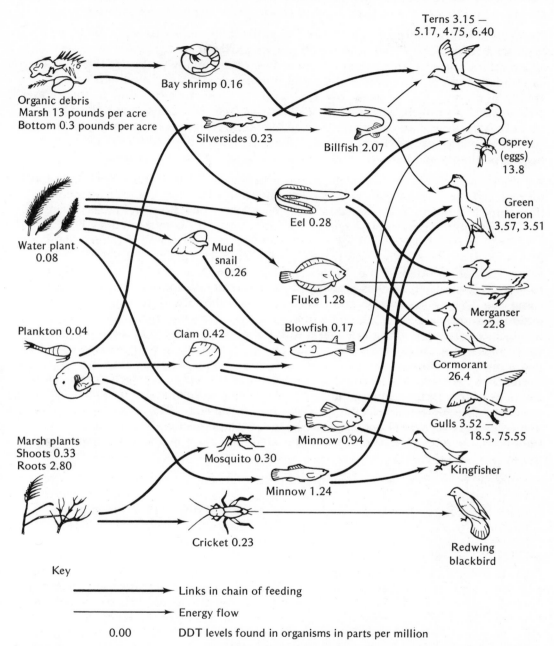

Terns 3.15 —
5.17, 4.75, 6.40

Bay shrimp 0.16

Organic debris
Marsh 13 pounds per acre
Bottom 0.3 pounds per acre

Silversides 0.23

Billfish 2.07

Osprey
(eggs)
13.8

Water plant
0.08

Eel 0.28

Mud
snail
0.26

Green
heron
3.57, 3.51

Fluke 1.28

Plankton 0.04

Clam 0.42

Blowfish 0.17

Merganser
22.8

Cormorant
26.4

Gulls 3.52 —
18.5, 75.55

Minnow 0.94

Marsh plants
Shoots 0.33
Roots 2.80

Mosquito 0.30

Minnow 1.24

Kingfisher

Cricket 0.23

Redwing
blackbird

Key

⟶ Links in chain of feeding

⟶ Energy flow

0.00   DDT levels found in organisms in parts per million

Figure 3.9   Pollution in a food web in the Long Island estuary, USA. Note the levels of DDT found in birds as compared with marsh plants.
Reprinted with permission from G. M. Woodwell, *Scientific American* 216, No 3 (1967) page 25, figure 1. Copyright © 1967 by Scientific American, Inc. All rights reserved

insect pests, but retention in animal tissue and consequent inimical or even fatal effects on animals at the top end of food chains make them serious pollutants.

The second reason pesticides are ineffectual in the long run is that many pests are evolving a resistance to chemicals which formerly killed them. For instance, insects which attack cotton plants have become resistant to so many pesticides that cotton can no longer be grown in parts of Central America, Mexico, and southern Texas. Thirdly, and perhaps surprisingly, the application of toxic chemicals may in some cases produce a pest problem where none previously existed. The Californian red scale (*Aonidiella aurantii*) is an insect which infests lemon trees but is naturally kept under control by many predators and parasites. Spraying lemon trees with DDT kills the natural predators and parasites and massive outbreaks of the red scale occur.

An alternative to pest control by pesticides is biological control. The general procedure for biological control runs like this. A pest, often one which has been introduced from outside a region, is causing severe damage. Insect predators and parasites are then sought in the pest's native region and introduced to the infested area. If successful, the pest population will be reduced to a level where no economic damage occurs. A case in point concerns the cactus known as the prickly pear. Native to North and South America, one species of prickly pear, *Opuntia stricta*, was brought to eastern Australia in 1839 as a hedge plant. It spread fast, forming dense stands, some one- to two-metres high and too thick for anyone to walk through. By 1900 it infested ten million acres. Control by poisoning the cactus was not economically feasible – clearing infested grazing land with poison was far more costly than the worth of the land. Searches in native habitats of *Opuntia stricta* were begun in 1912 to find a possible biological control agent. Eventually one was found – a moth, *Cactoblastis cactorum*, native to northern Argentina, the larvae of which burrow into the pads of the cactus, causing damage and introducing bacterial and fungal infections. The moth was introduced to eastern Australia in 1925. Success was almost immediate. Between 1930 and 1931, when the moth population had become enormous, the prickly pear stands were ravaged. By 1940, the prickly pear was still found here and there, but nowhere was it a pest.

Another alternative to chemical control of pests is genetic control. One way of doing this is to breed crop plants which are more resistant to pests. It was this method of control that came to the rescue of the French wine industry in 1891. Phylloxera is an aphid native to America which lives in galls on leaves and roots of vines where it cannot be reached by sprays. It multiplies prodigiously. Vines infected by it become stunted and die. It was accidentally introduced to the Languedoc region of France in 1861. Two decades later, four-fifths of the Languedoc vineyards had been devastated and every vine-growing area of France was infected; no remedy had been found; the outlook was bleak for all wine drinkers. In 1891, it was discovered that the American vine (*Vitis labrusca*) was almost immune to phylloxera. Scions of the European vine (*Vitis vinifera*) were grafted onto American root stocks to produce a hybrid vine which, if not entirely immune, was affected far less seriously.

Further Reading

Geography: A Modern Synthesis (third edition), P. Haggett, Harper & Row (1979), Chapter 8.
Geography and Man's Environment, A. N. Strahler et al., Wiley (1977), Part V.